YOUR JOURNEY STARTS HERE

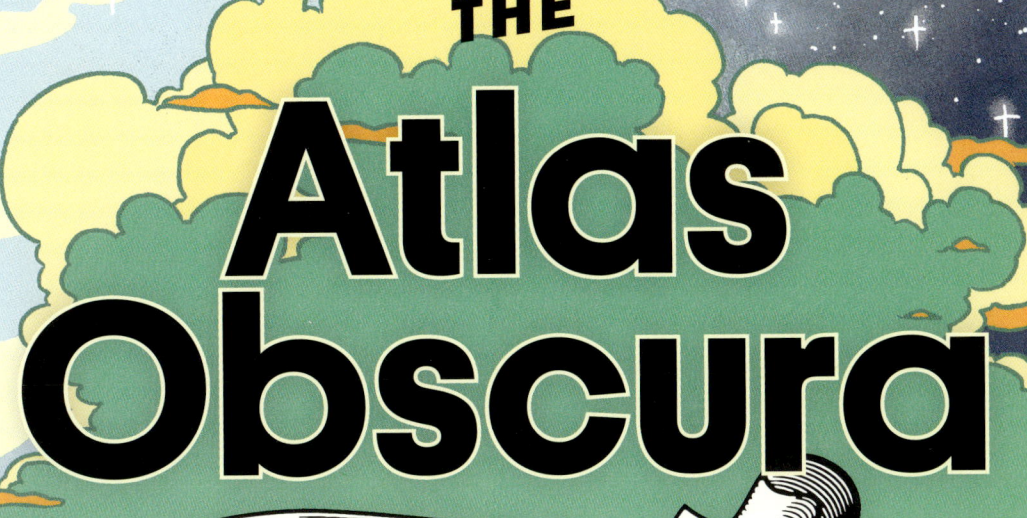

THE
Atlas Obscura

EXPLORER'S GUIDE

TO

INVENTING THE WORLD

Dylan Thuras and
Jennifer Swanson

Illustrated by Ruby Fresson

WORKMAN PUBLISHING • NEW YORK

ASTRONOMY P. 32

AIRPLANES P. 52

For Jean, Phineas, and Michelle, I love you all dearly.
For James Burke, the great connector, for being a delight
and an inspiration. For children everywhere,
the future is yours to invent.
—DYLAN THURAS

To all of the curious kids, like me, who were always
asking questions and wondering how things worked.
—JENNIFER SWANSON

For Ginger, Bringer of Chaos
—RUBY FRESSON

STEEL P. 66

SEAFARING P. 60

GUNPOWDER P. 48

Workman Kids
Workman Publishing
Hachette Book Group, Inc.
1290 Avenue of the Americas
New York, NY 10104
workman.com

Workman Kids is an imprint of Workman Publishing, a division of Hachette Book Group, Inc.
The Workman name and logo are registered trademarks of Hachette Book Group, Inc.

Illustrations by Ruby Fresson • Design by Sara Corbett

Library of Congress Cataloging-in-Publication Data is available.
ISBN 978-1-5235-1688-9
First Edition August 2025 APS

Printed in Penang, Malaysia,
on responsibly sourced paper.

10 9 8 7 6 5 4 3 2 1

ELECTRICITY P. 72

MAPS P. 58

FIRE P. 6

*Inventing the World is a wondrous journey through time, space, and science.
It is filled entirely with real places, many of which are very much worth visiting. Some places
in this book, however, are not open to the public and are not meant to be visited without
appropriate permissions, or even at all! Travel with respect, obey the local laws, and don't go
to places that are dangerous or off-limits. Lastly, the lawyers want you to know that "Neither
the author nor the publisher shall be liable or responsible for any loss, injury, or damage
allegedly arising from any information or suggestions contained in this book." OK, let's explore!*

HELLO, INVENTORS!

What is technology?

Is it lasers? Is it robots? Is it robots with laser vision? Yes, all those things are technology, but so is the chair you are sitting on, the paper this book is printed on, and even the alphabet itself. We are fish living in an ocean of invention!

Look around you: Maybe there is a lamp nearby. To create that lamp, you would first need to learn how to control fire, then find a way to use that fire to melt metal out of rocks, and then turn sand into glass. Good—now we just need to discover electricity, build a motor that generates it, discover what materials emit light when electricity passes through them, and on and on and on. Our entire human-built world of roads, buildings, farms, energy, and so much more is all linked through a colossal, interconnected web of invention reaching back tens of thousands of years.

Of course, technology is also capable of harm. After all, humans burned fossil fuels and ended up heating our atmosphere, creating climate change. We are now racing to build new technologies, better solar panels, electric cars, and more to help fix our mistakes. It's up to societies to choose how we use our brilliance. But making wise decisions means *understanding* the inventions and technologies that shape our world. We must become fish that can *see* the water! So let's dive in!

You are about to travel through 50 inventions and discoveries, each one connecting to the next, back and forth through time and around the world. Although the relationship between inventions is rarely simple, this book charts one path along the incredible web of technology. These 50 ideas represent some of the most important inventions of our past—and some that will shape our future.

Let's go invent the world.

PACKING LIST

On this journey through time and invention, we'll be visiting ancient civilizations, dangerous machines, and laboratories experimenting with some of the most extreme energies on Earth. We'll peer deep into the microscopic world and look out across the vast universe. To pack for a trip like this, we'll need good gear—and it's possible that we just might have to invent a few new things ourselves . . .

TIME MACHINE

First things first. Since we are going on a trip that takes us back tens of thousands of years, we will need a time machine. Technically, a time machine violates laws of physics and so is impossible to build . . . But who cares! We've got one!

LEAD APRON

It's not just heat we have to worry about; we are also going to be working with some high-energy waves like X-rays. Best to bring our own lead apron just in case. (Although if you decided to stand in front of the X-Ray Free Electron Laser . . . good luck, because this apron won't protect you.)

MOTION SICKNESS TABLETS

We will be traveling via some unusual transportation on this journey as we visit space shuttles, ancient ships, and high-speed trains. It's probably a good idea to pack something to settle the stomach.

HEAT RESISTANT SUIT

We will be working with fire, molten metal, and even plasma. An aluminized suit— basically a supershiny set of pajamas made of aluminum foil—with fire-resistant fabrics underneath seems like a good idea!

UNIVERSAL TRANSLATOR

We'll be traveling all around the world and back in time, so we'll need a way to communicate. A universal translator that understands all languages will be very useful. Good thing you invented it!

χαῖρε

HI!

LASER SAFETY GLASSES

Speaking of lasers, we'll soon be around a *lot* of them: X-ray lasers, lasers in ultraprecise clocks, and AI-controlled lasers that zap weeds. So we're gonna need some glass to keep your eyes protected.

A NOTEBOOK

Compared to other tech, a notebook might seem a bit boring, but this is a must! Much of what we are exploring involves a lot of complicated science and math, and a notebook is the place to write all your questions. Maybe even try your hand at some codebreaking!

```
20  8   5
 5  14  5   16
13  9   19
20  9   7   8
20  8   5   18
20  8   1   14
20  8   5   14
23  15  18  19
15  18  4
```

SAMURAI SWORD

Trust us, it's always good to have one on hand.

FOX TREATS

A nice mix of cherries, apples, and worms for any adorable foxes you meet . . .

TOWEL

Use it to keep warm in freezing labs or cover your face in the sandy ruins of ancient Rome. It's about the most massively useful thing a world inventor can have.

INVENTION & DISCOVERY GUIDE

★ On each page, next to the name of the city, town, or region, you'll find a set of numbers and letters that look like this: Ⓝ 60.9025 Ⓔ 101.9045. These are your destination's *latitude* and *longitude*. What do they mean? They're a kind of special code. Picture a huge grid of lines crisscrossing the surface of the Earth. Lines of latitude stretch from east to west, and lines of longitude run from the North Pole to the South Pole. The numbers and letters give you the exact coordinates of your destination on this grid. Here's the coolest part: If you type the coordinates into Google Maps (or another online map service), it'll take you right there!

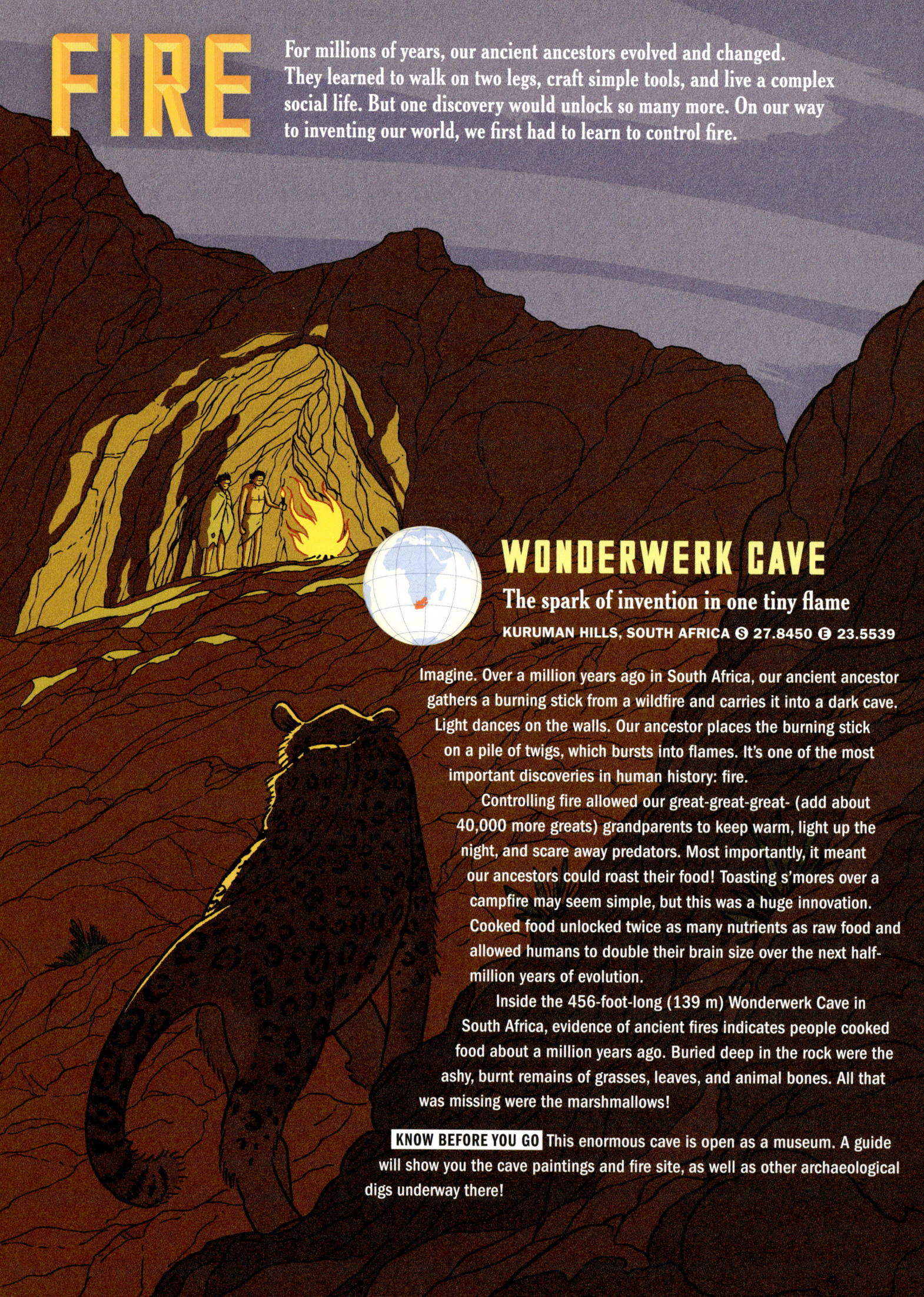

FIRE

For millions of years, our ancient ancestors evolved and changed. They learned to walk on two legs, craft simple tools, and live a complex social life. But one discovery would unlock so many more. On our way to inventing our world, we first had to learn to control fire.

WONDERWERK CAVE

The spark of invention in one tiny flame

KURUMAN HILLS, SOUTH AFRICA Ⓢ 27.8450 Ⓔ 23.5539

Imagine. Over a million years ago in South Africa, our ancient ancestor gathers a burning stick from a wildfire and carries it into a dark cave. Light dances on the walls. Our ancestor places the burning stick on a pile of twigs, which bursts into flames. It's one of the most important discoveries in human history: fire.

Controlling fire allowed our great-great-great- (add about 40,000 more greats) grandparents to keep warm, light up the night, and scare away predators. Most importantly, it meant our ancestors could roast their food! Toasting s'mores over a campfire may seem simple, but this was a huge innovation. Cooked food unlocked twice as many nutrients as raw food and allowed humans to double their brain size over the next half-million years of evolution.

Inside the 456-foot-long (139 m) Wonderwerk Cave in South Africa, evidence of ancient fires indicates people cooked food about a million years ago. Buried deep in the rock were the ashy, burnt remains of grasses, leaves, and animal bones. All that was missing were the marshmallows!

KNOW BEFORE YOU GO This enormous cave is open as a museum. A guide will show you the cave paintings and fire site, as well as other archaeological digs underway there!

FIRE RESEARCH LABORATORY

Burn to learn

BELTSVILLE, MARYLAND, USA Ⓝ 39.0570 Ⓦ 76.9009

Just north of Washington, DC, there is an ordinary-looking house. Maybe it's similar to yours. It has furniture, books, and food in the kitchen cupboards. But this house is on fire. Luckily, the house has been specially built inside a massive, 16,900-square-foot (1,570 m²) "burn room" at the US government's Fire Research Laboratory. For the Certified Fire Investigators who work in the lab, setting stuff ablaze is their job. Why? By simulating fires, these researchers learn how fires start, how they spread, and even how to tell if they're accidental or set on purpose.

Other fire investigation labs experiment with fire scenarios from nature. By creating mini "fire tornadoes," they can study how the real-world ones work. A fire tornado is a whirling vortex of flame that can form during large fires. These "firenadoes" can reach 1,000 feet (305 m) in diameter, burn at 2,000°F (1,093°C), and produce winds of 160 mph (257 kph). Even though controlling fire was the first great invention of humankind, we still have much to learn.

KNOW BEFORE YOU GO The fire lab is the only place in the country dedicated to investigating criminal fires. The Bureau of Alcohol, Tobacco, Firearms and Explosives, which operates the lab, also offers forensic science internships for high school students.

GLASS

An ancient Roman story goes like this: Thousands of years ago a group of sailors were shipping a mineral called natron. They made a campfire on the beach and propped up their fiery cauldron with natron chunks. As the natron and beach sand heated up together, the sailors "beheld transparent streams flowing forth." This, said the story, "was the origin of glass."

ZHANGJIAJIE GLASS BRIDGE

Don't look down!

WULINGYUAN SCENIC AREA, ZHANGJIAJIE, CHINA Ⓝ 29.3987 Ⓔ 110.6982

Step out on the Zhangjiajie Glass Bridge and try not to panic. Sure, you're standing on a sheet of glass. Sure, you can watch birds flying *beneath* your feet in the 984-foot-deep (300 m) chasm below. But don't worry. Just because it's made of glass doesn't mean it's not strong. Spanning more than 1,400 feet (427 m) across the Zhangjiajie Grand Canyon, this see-through bridge provides a literal bird's-eye view of the valley's gorgeous forests, waterfalls, and streams.

Four steel pillars support a horizontal steel frame that holds 120 panels of 2-inch-thick (5 cm) triple-layered glass. Having been heated to a temperature of 1,292°F (700°C), then cooled quickly, the glass becomes "tempered," which increases its strength. Tempered glass is four times harder to break than regular glass, which is a very good thing to keep in mind as you walk (or crawl!) across the bridge.

KNOW BEFORE YOU GO To visit the bridge you must take a cable car to the top. We recommend against crossing if you are afraid of heights. Seems wise.

BENI SALAMA

A transparently amazing early Roman glassmaking factory

WADI EL NATRUN, EGYPT Ⓝ 30.3249 Ⓔ 30.3249

As the wind blows across a hot desert valley, a man in a simple tunic stands beside a massive furnace full of hot bright-orange liquid. He is busy creating one of the most exciting materials in the ancient world: glass.

This desert in Wadi El Natrun, Egypt, is filled with a special salt called natron. The ancient Egyptians used it to dry mummies, make soap, and treat wounds. But natron had another amazing use: making glass. Mix sand and natron salt, then heat it all to 2,012°F (1,100°C), and voilà—you have molten glass.

The ready supply of natron led the Romans to build large glass factories here 2,000 years ago. Using a blowpipe, glassblowers could blow the molten glass into useful shapes. Until then, glass was an expensive material used in things like jewelry. But Roman factories like Beni Salama transformed it into something everyday people could afford. Today, glass is everywhere—on our buildings, in our screens, and even under our feet.

KNOW BEFORE YOU GO Beni Salama dates back much further than the Romans to 4800 BCE and is the earliest North African archaeological site showing plant and animal domestication. It also includes a staircase of hippopotamus bones!

LENSES

You might not think about yourself as having "lenses." But believe it or not, the natural lenses in your eyes are what allow you to read this book. Thousands of years ago, when humans first shaped pieces of glass to focus light, these artificial lenses turned out to be as useful in our hands as the ones in our heads.

VERA C. RUBIN TELESCOPE

The world's largest lens can see for trillions of miles.

ELQUI PROVINCE, CHILE Ⓢ 30.2446 Ⓦ 70.7494

At the top of an 8,900-foot-high (2,712 m) mountain in northern Chile, you can peer through a lens as large as a person. At 5 feet (1.5 m) across, the Vera C. Rubin telescope (named after a famous American astronomer) holds some of the largest glass lenses ever created.

Why so big? Well, this telescope is basically an enormous camera. Huge mirrors gather the light of objects millions of light-years away, and three humongous lenses help focus the light for the giant camera. This telescope detects so much light that it can see supergiant stars, extremely distant planets, and even black holes, like one at the center of the Messier 87 galaxy, approximately 54 million light-years away from us.

The Rubin telescope is large, and it's also fast! It rotates in every direction and takes 1,000 panoramic images of the night sky each day. Scientists plan to use those images to create a recording of the cosmos and a 3D map of the universe. Imagine watching a star collapse—talk about an eye-opening experience!

KNOW BEFORE YOU GO The Vera C. Rubin Observatory is reachable only by dangerous mountain roads, so you'll need a trained driver to get up there!

THE NIMRUD LENS

The world's oldest lens— or just a bit of bling?

BRITISH MUSEUM, LONDON, UK
Ⓝ 51.5194 Ⓦ 0.1269

Buried in an ancient palace, surrounded by magnificently carved stone panels of Assyrian kings, a seemingly unremarkable piece of rock crystal was discovered. It wouldn't be special at all, except for one thing: It magnifies the world.

Light slows down when it goes through water or glass. It also changes direction based on the angle at which it hits those materials. With a curved lens (like the one you'd find in a magnifying glass), you can focus many rays of light down to a single dot, called a focal point. When you hold the lens so the focal point is at your eye, the bending light makes the object appear larger.

Now residing in the British Museum, the 2,700-year-old Nimrud Lens isn't especially powerful—it only makes things look three times bigger. But it could be the earliest example of lens power. Or maybe it was just shiny. Some archaeologists say it was used as decoration on furniture. We may never know, but many ancient peoples certainly understood that by forming glass or crystal into the right shape, they could control light itself.

KNOW BEFORE YOU GO The lens was discovered in modern-day Iraq and is found in the museum's Assyrian collections. While there, keep an eye out for some of the weirder items, like the crystal skull or the mummified cats!

MICROSCOPES

Lenses help us see our everyday world, the skies above us, and so much more. In the 1600s, Dutch eyeglass makers created lenses that let them magnify objects, making them tens of times larger than normal. These instruments, called mikrós skopéō (microscopes), even revealed the tiny worlds inside us!

HOOKE'S MICROSCOPE

Illuminating an invisible world

NATIONAL MUSEUM OF HEALTH AND MEDICINE, SILVER SPRING, MARYLAND, USA Ⓝ 39.0087 Ⓦ 77.0538

Look at your hand. What do you see? Skin. Fingernails. Wrinkles around your knuckles, most likely. But what if you could see it much, much closer? What would be there? At America's official medical museum near Washington, DC, you will find a 7-inch-long (17.8 cm) tube made of cardboard and leather that is a portal to this invisible world.

In the 1660s, scientist Robert Hooke experimented with a microscope invented by the famous astronomer Galileo Galilei. He realized that by adjusting the positions of the three different lenses within and using good lamps, he could magnify objects, making their parts appear many times larger than normal. With his improved microscope, Hooke studied a piece of tree cork and observed that it was made of tiny compartments. He named them "cells" after the little rooms that monks lived in. Hooke had discovered the fundamental unit of life that makes up all living things. We still call them cells to this day!

KNOW BEFORE YOU GO The museum also features bones, brains, skulls, and many other amazing medical specimens. Keep an eye out for the bullet that killed President Lincoln.

MESOLENS

A massive microscope that views tiny things in new ways

UNIVERSITY OF STRATHCLYDE, GLASGOW, SCOTLAND Ⓝ 55.8629 Ⓦ 4.2417

Hooke's microscope could view a single hair on a flea, but the Mesolens microscope can look at *every* hair on the entire flea and even see the cells inside the hairs. With an extremely large lens system—about the length and width of a human arm—the Mesolens is uniquely powerful.

Designed by Professor Gail McConnell and Dr. Brad Amos, the Mesolens can give a high-definition, three-dimensional view of the specimen. That means scientists can see different depths of the sample. Using bright light, colored dyes, and a moving focus, the Mesolens can even look inside a specimen layer by layer. Typical microscopes can't do all that.

The Mesolens allows scientists to see into the guts, brain, chest, and bacteria inside an organism, all in one picture. Researchers can study the dynamic relationships between bacteria rather than look at them one by one. This microscope's penetrating vision was recently used to make detailed images of infected tonsils and will potentially help kids avoid tonsil surgery!

KNOW BEFORE YOU GO Only certain scientists are permitted to visit the Mesolens, but anyone can tour the university. Using the Mesolens website, you can also see some of the amazing images the microscope creates.

MICROBES

You know those tiny organisms that scientists discovered under the microscope? Well, every person has about 38 trillion of these single-celled life-forms (called microbes) living inside them. In fact, more than half of the cells in your body aren't even human! But don't worry, they're just different species of bacteria, as well as fungi like yeast. Still sound bad? Turns out you need them just as much as they need you!

MICROPIA

The world's first microbe zoo

AMSTERDAM, NETHERLANDS Ⓝ 52.3669 Ⓔ 4.9126

Imagine a planet that has trillions of inhabitants. These strange creatures wiggle, squiggle, and multiply. They are some of the most powerful organisms known to exist in nature. That planet is you!

At Micropia, step right up to the body scanner, which uses a camera to zoom in on different body parts. You'll discover the many strange creatures that call your eyes, nose, mouth, hands, feet, hair, and skin "home."

Meet the microbes that live in your guts and help you digest food. Or microbes that send signals to your brain and help you feel happy or sad. There are scary microbial "monsters" out there too, like parasitic fungi, which enter the brains of spiders and turn them into spider "zombies." You'll even get a close-up view of the microbes that help create vaccines, as well as some of the world's oldest extremophiles—microbes that date back to the beginning of the Earth.

KNOW BEFORE YOU GO Micropia is part of the Artis Zoo complex, which includes Amsterdam's oldest zoo, a planetarium, an aquarium, and a science museum with dozens of skeletons and interactive exhibits about the body.

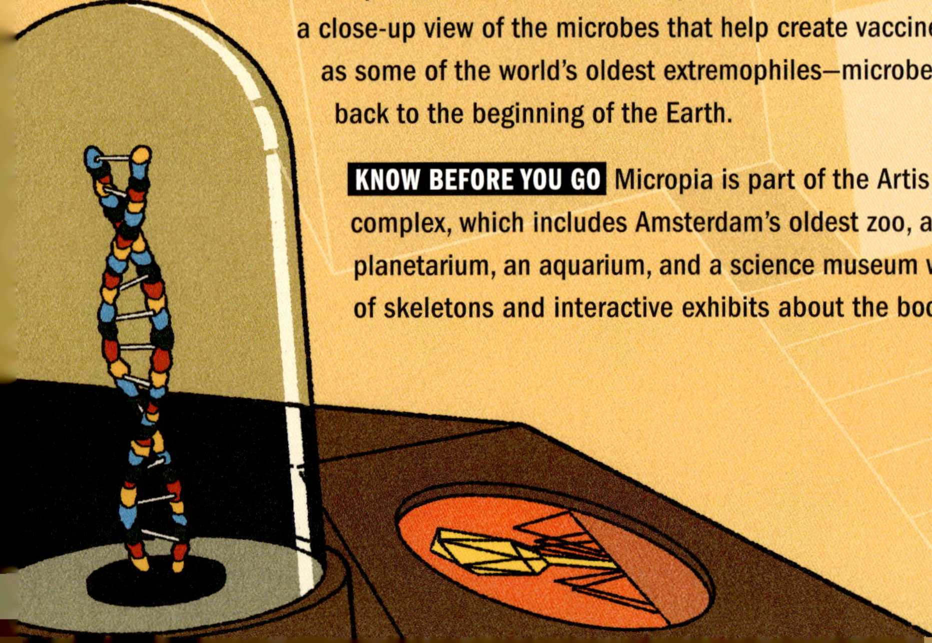

PASTEUR MUSEUM

The home that germinated germ theory

PARIS, FRANCE Ⓝ 48.8403 Ⓔ 2.3114

Among the antique chairs and sofas of a beautifully decorated apartment sits a rack of empty test tubes, flasks, and microscopes untouched for years. This home belonged to the famous scientist Louis Pasteur, the man who helped discover microorganisms (microbes) that make us sick.

In the early 1800s, people believed they fell ill because they breathed in "foul odors" or "bad air." But Pasteur had a different theory. He identified tiny microbial bacteria—what he called germs—that got into wine and beer. He realized that it was the germs that caused illness.

Putting his theory into practice, Pasteur invented a system for eliminating bacteria in liquids by heating them up. Cooks in China and Japan had been doing this for centuries, but Pasteur scientifically proved why it worked: The heat kills the germs, making the liquids safe to drink. We still use this method to "pasteurize" milk, as well as juice, almonds, and even ice cream! Pasteur's theory of germs has saved countless lives.

KNOW BEFORE YOU GO Besides the incredible antique science equipment, you can also visit the enormous crypt where Pasteur and his wife, Marie, are buried.

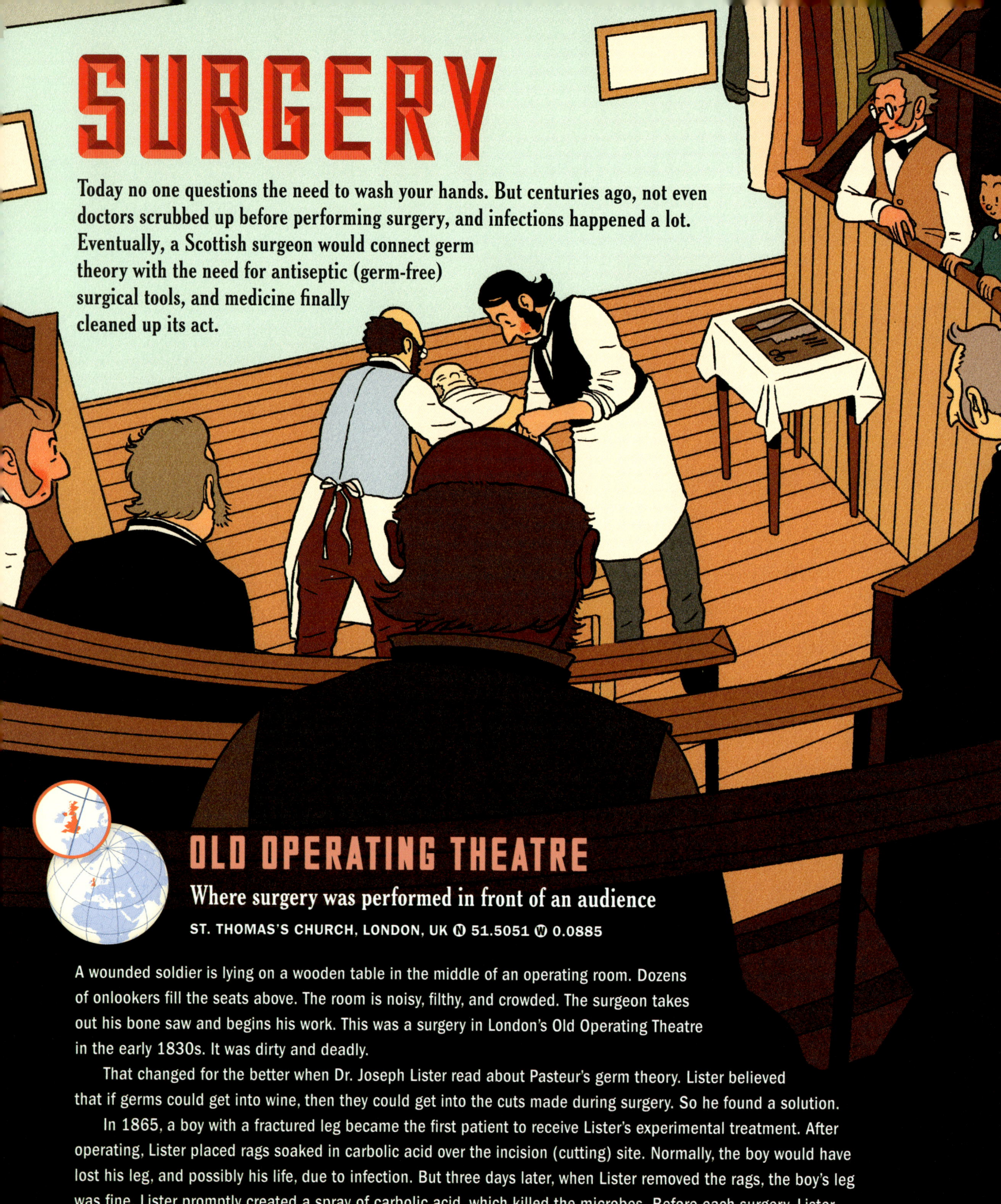

SURGERY

Today no one questions the need to wash your hands. But centuries ago, not even doctors scrubbed up before performing surgery, and infections happened a lot. Eventually, a Scottish surgeon would connect germ theory with the need for antiseptic (germ-free) surgical tools, and medicine finally cleaned up its act.

OLD OPERATING THEATRE

Where surgery was performed in front of an audience

ST. THOMAS'S CHURCH, LONDON, UK Ⓝ 51.5051 Ⓦ 0.0885

A wounded soldier is lying on a wooden table in the middle of an operating room. Dozens of onlookers fill the seats above. The room is noisy, filthy, and crowded. The surgeon takes out his bone saw and begins his work. This was a surgery in London's Old Operating Theatre in the early 1830s. It was dirty and deadly.

That changed for the better when Dr. Joseph Lister read about Pasteur's germ theory. Lister believed that if germs could get into wine, then they could get into the cuts made during surgery. So he found a solution.

In 1865, a boy with a fractured leg became the first patient to receive Lister's experimental treatment. After operating, Lister placed rags soaked in carbolic acid over the incision (cutting) site. Normally, the boy would have lost his leg, and possibly his life, due to infection. But three days later, when Lister removed the rags, the boy's leg was fine. Lister promptly created a spray of carbolic acid, which killed the microbes. Before each surgery, Lister sprayed it on his tools, bandages, and thankfully, even his hands.

KNOW BEFORE YOU GO The Old Operating Theatre gives you a glimpse back to the deadly old days before Lister's experimental surgery. To reach it, you'll climb up a really cool old spiral staircase.

THE DA VINCI ROBOTIC SURGICAL SYSTEM

The future of surgery is robotic.

SURGEONS' HALL MUSEUMS, EDINBURGH, SCOTLAND Ⓝ 55.9467 Ⓦ 3.1853

If the invention of antiseptic surgery was a huge leap forward, the giant robot in Scotland's Surgeons' Hall Museums may be the next big jump.

A normal operation involves a surgeon cutting into a patient with a scalpel. But in a person's midsection, where organs sit very close together, there isn't a lot of space to work. Enter: the da Vinci robotic surgical systems.

Standing 5 feet, 9 inches (1.75 m) tall and 3 feet (0.9 m) wide, and weighing more than 1,200 pounds (544 kg), the massive da Vinci machine is the first thing you'll notice when you walk into the Body Voyager exhibit. Its four moveable arms become the surgeon's hands. During an operation, the surgeon watches on a screen while maneuvering the robot's microscopic tools. Da Vinci robots have performed more than 10 million surgeries and help improve recovery time in most cases, but they're not foolproof. They're only as good as the surgeons operating them!

KNOW BEFORE YOU GO The museums have several floors filled with surgical specimens (body parts) in jars, as well as collections related to crime and dentistry. Visitors can even operate the da Vinci robot!

GENETICS

Before surgeons wrapped wounds with medical gauze, they used simple rags. These pus rags (yes, we said "pus rags") would later provide the material for one of the first great breakthroughs in understanding the very code of life itself: genetics.

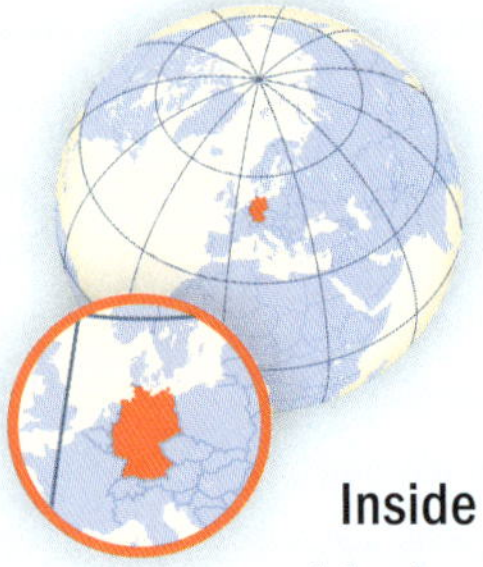

SCHLOSSLABOR LABORATORY

The building blocks of life, discovered in a medieval kitchen

HOHENTÜBINGEN CASTLE, TÜBINGEN, GERMANY Ⓝ 48.5192 Ⓔ 9.0503

Inside a medieval castle in Germany is a chemical lab in a 500-year-old kitchen. It was here that the building blocks of life were first discovered.

In the 1800s, a partially deaf scientist named Friedrich Miescher wanted to study white blood cells, which were believed to help with healing. But these cells were not easy to obtain, especially since they had to be extracted from a patient's blood. What to do?

Miescher had a terrific (and horrifying) realization. The pus rags—bandages for the yellowish goo a wound creates when it's fighting an infection—discarded during surgery were chock-full of white blood cells! So he gathered up some rags, scraped off the pus, and isolated one particular molecule that seemed to make up much of the white blood cells. At first, Miescher didn't understand what he had found. But the substance he discovered, nucleic acid, is the basis of DNA—the molecule that carries genetics, or the basic code of life. The medieval castle kitchen became known as the "cradle of biochemistry," and pus rags were never looked at the same way again.

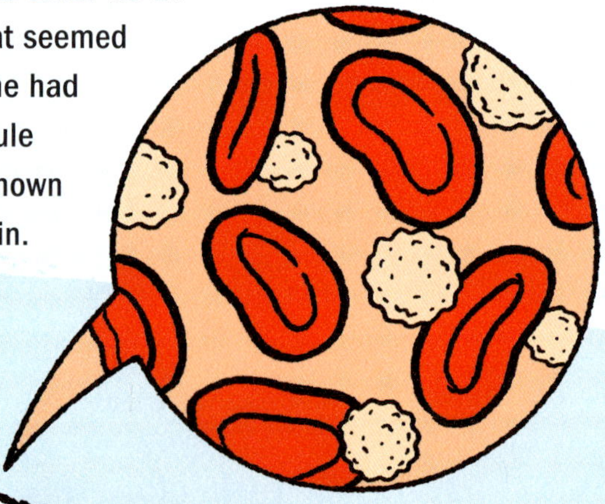

KNOW BEFORE YOU GO Hohentübingen Castle is home to a wonderful archaeological collection, but if you're looking for the room that made pus rags famous, head downstairs to the castle laboratory.

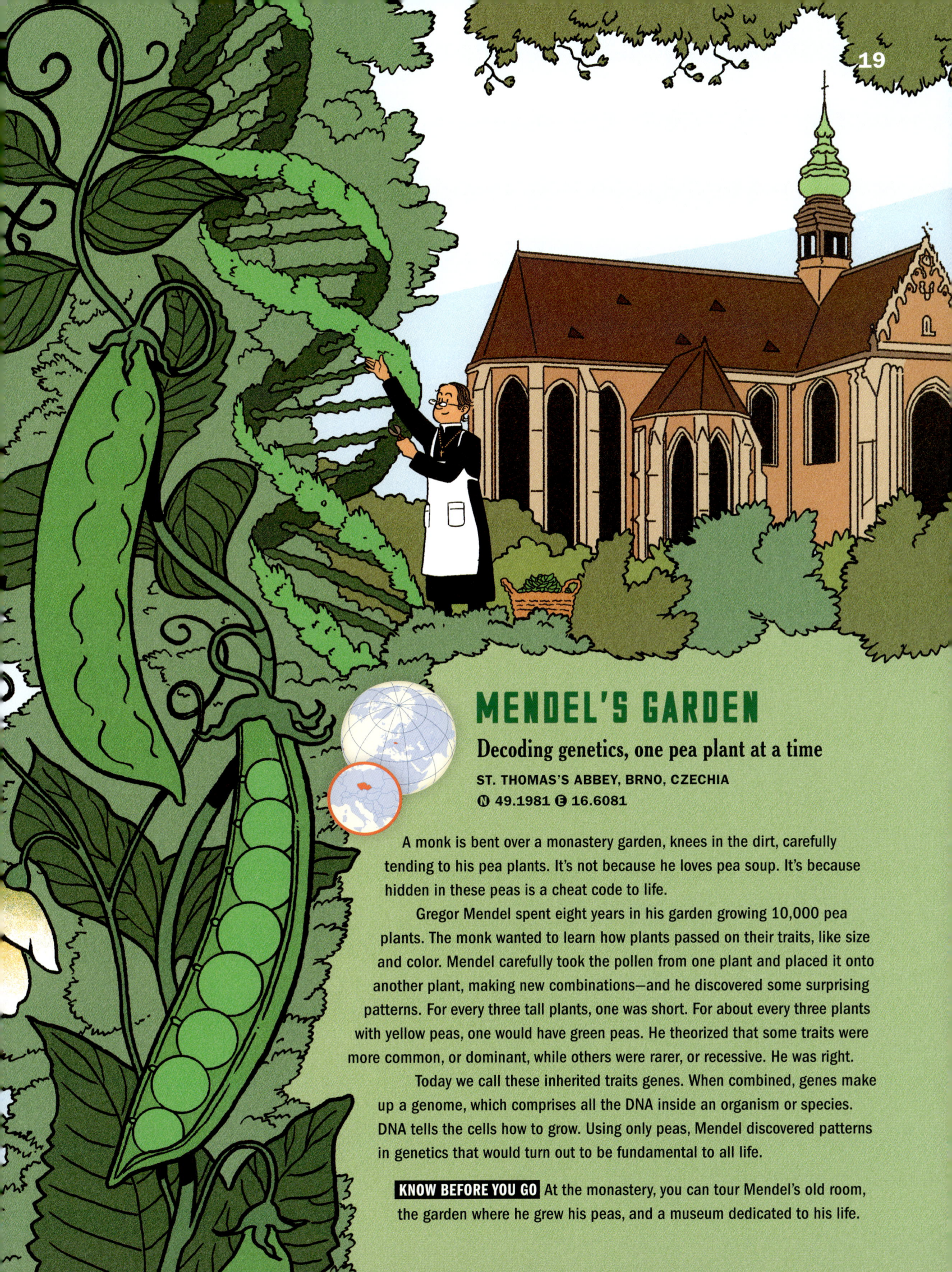

MENDEL'S GARDEN

Decoding genetics, one pea plant at a time

ST. THOMAS'S ABBEY, BRNO, CZECHIA
Ⓝ 49.1981 Ⓔ 16.6081

A monk is bent over a monastery garden, knees in the dirt, carefully tending to his pea plants. It's not because he loves pea soup. It's because hidden in these peas is a cheat code to life.

Gregor Mendel spent eight years in his garden growing 10,000 pea plants. The monk wanted to learn how plants passed on their traits, like size and color. Mendel carefully took the pollen from one plant and placed it onto another plant, making new combinations—and he discovered some surprising patterns. For every three tall plants, one was short. For about every three plants with yellow peas, one would have green peas. He theorized that some traits were more common, or dominant, while others were rarer, or recessive. He was right.

Today we call these inherited traits genes. When combined, genes make up a genome, which comprises all the DNA inside an organism or species. DNA tells the cells how to grow. Using only peas, Mendel discovered patterns in genetics that would turn out to be fundamental to all life.

KNOW BEFORE YOU GO At the monastery, you can tour Mendel's old room, the garden where he grew his peas, and a museum dedicated to his life.

DOMESTICATION

Humans had been practicing their own informal versions of Mendel's garden for thousands of years. The breeding of plants and animals, or adapting wild versions to more useful ones, was one of the earliest human technologies. For more than 10,000 years, from lima beans to llamas, people have guided the natural world all around them through domestication.

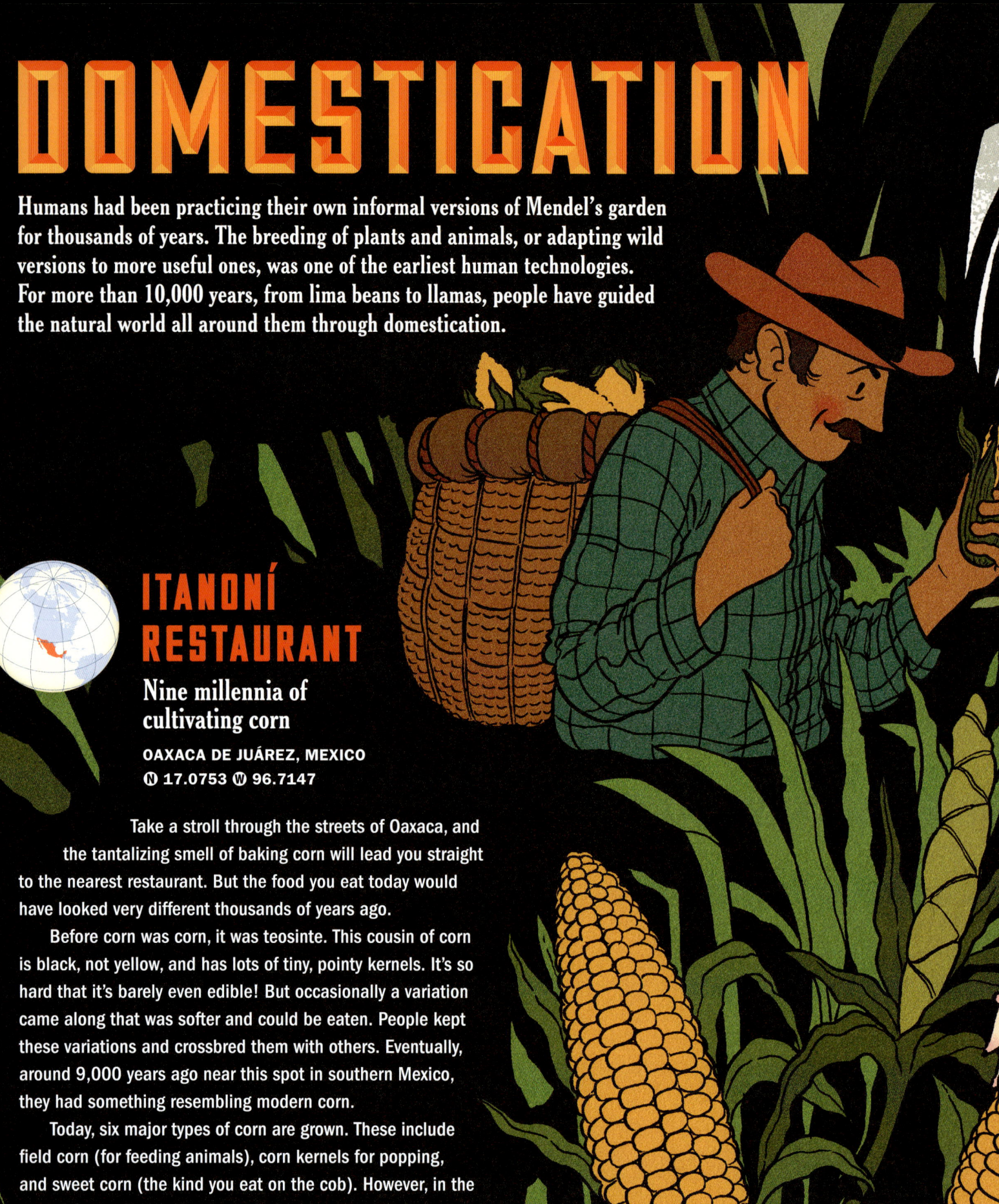

ITANONÍ RESTAURANT

Nine millennia of cultivating corn

OAXACA DE JUÁREZ, MEXICO
Ⓝ 17.0753 Ⓦ 96.7147

Take a stroll through the streets of Oaxaca, and the tantalizing smell of baking corn will lead you straight to the nearest restaurant. But the food you eat today would have looked very different thousands of years ago.

Before corn was corn, it was teosinte. This cousin of corn is black, not yellow, and has lots of tiny, pointy kernels. It's so hard that it's barely even edible! But occasionally a variation came along that was softer and could be eaten. People kept these variations and crossbred them with others. Eventually, around 9,000 years ago near this spot in southern Mexico, they had something resembling modern corn.

Today, six major types of corn are grown. These include field corn (for feeding animals), corn kernels for popping, and sweet corn (the kind you eat on the cob). However, in the search for perfect, easy-to-grow corn, we have lost many of the ancient, diverse varieties. Itanoní restaurant keeps ancient traditions alive by serving many delicious local varieties.

KNOW BEFORE YOU GO Itanoní specialties include a sweet corn drink called atole de panela and toasted triangular tortillas filled with cheese. Sit near the clay griddle to see the cooks at work.

THE DOMESTICATED SILVER FOX EXPERIMENT

An experimental program to make foxes friendly

FARM NEAR THE INSTITUTE OF CYTOLOGY AND GENETICS, NOVOSIBIRSK, RUSSIA Ⓝ 54.8480 Ⓔ 83.1067

There are few things better than petting your dog. But dogs weren't always friendly—in fact, they weren't always dogs! Their ancient ancestors were wild wolves, and in many places humans feared and hunted wolves. So, how is it that 27,000 years later, you have a dog that loves to cuddle?

It may have started with a wolf that was just a bit friendlier, liked eating leftovers from humans, and stuck around the camp. People learned the advantages of having a wolf to help chase down prey on a hunt. The wolves that were good at cooperating with humans survived and bred with each other, and bit by bit they changed. Over many generations, they evolved tails that can wag, floppy ears, and a love of petting.

The longest-running study on animal domestication, started in 1959 and now headed by geneticist Lyudmila Trut, does the same thing with silver foxes by breeding the calmest and friendliest foxes. Over generations, the foxes began to change in the way dogs had long ago. The foxes have evolved more rounded bodies and short, curly tails, and they are much more playful. Cuteness is a powerful evolutionary strategy!

KNOW BEFORE YOU GO While the farm is for researchers only, an internet search will bring up pictures of the foxes they are breeding. We dare you not to say "*awwww*."

ANCIENT AGRICULTURE

Domesticating plants and animals led to farming—and farming changed everything. Like many discoveries, agriculture came with both good and bad. Farms allowed people to stay in one place, have bigger families, and make more food. But people also worked longer hours, and diseases spread among settlements. Bit by bit, farming took over the ancient world.

TERRACED FARMS OF THE ANCIENT INCAS

A stepping stone to incredible eating

MORAY MARAS, PERU Ⓢ 13.3292 Ⓦ 72.1956

Picture a farm. Maybe you see a vast field of grain. Try again.

If you're an Incan living more than 800 years ago in Moray, Peru, a farm looks like several giant pits dug out of the ground, with enormous circular "steps" cut into the sides, like an upside-down wedding cake. These pits, now known as the Moray ruins, were more than just a farm. They were an agricultural laboratory.

Incan farmers didn't have a lot of flat land to plant their crops, so they got creative by carving huge, flat steps into the sides of the mountain, a technique called terrace farming. The steps added growing space, and their stone retaining walls absorbed sunlight and heat. The bottoms were mixed with gravel that drained slowly and conserved water. It's believed that at Moray, the Inca experimented with different terrace techniques, trying out many soil and growing conditions.

Terrace farming is so efficient that modern Peruvian farmers are returning to it today. Sometimes a step back in time is a step in the right direction. ("Steps"—get it? You get it.)

KNOW BEFORE YOU GO Overlooking the ruins is a restaurant called Mil that uses Incan ingredients that might once have been grown at Moray. It's been named one of the 50 best restaurants in the world, so make reservations!

THE KAREZ WELL SYSTEM

The Great Well of China

TURPAN, CHINA Ⓝ 42.8730 Ⓔ 89.3980

Plants need healthy seeds, good land, and plenty of water, of course. But where do you find water if you're farming in the desert?

The Uyghur people in the Xinjiang region of China built an incredible underground irrigation (watering) system. Known as the Karez well system, it captures melting water from glaciers far away in the mountains and funnels the water for hundreds of miles to the dry desert city of Turpan.

The Uyghur began digging these wells by hand more than 2,000 years ago. At its height, the system covered more than 3,100 miles (4,989 km), which is longer than the distance from New York City to Los Angeles. Every 60 to 230 feet (18 to 70 m) along the tunnels was a hole with a bucket and pulley system. Farmers could pull water up and use it to irrigate their crops. Despite their age, a few hundred of these wells are still in use today.

KNOW BEFORE YOU GO The Turpan Karez Museum offers trips through the well tunnels, though at the moment trips to the region aren't recommended due to oppression of the Uyghur by the Chinese government.

MOUNTAIN

WELLS

FARMLAND

BEDROCK

AQUIFER

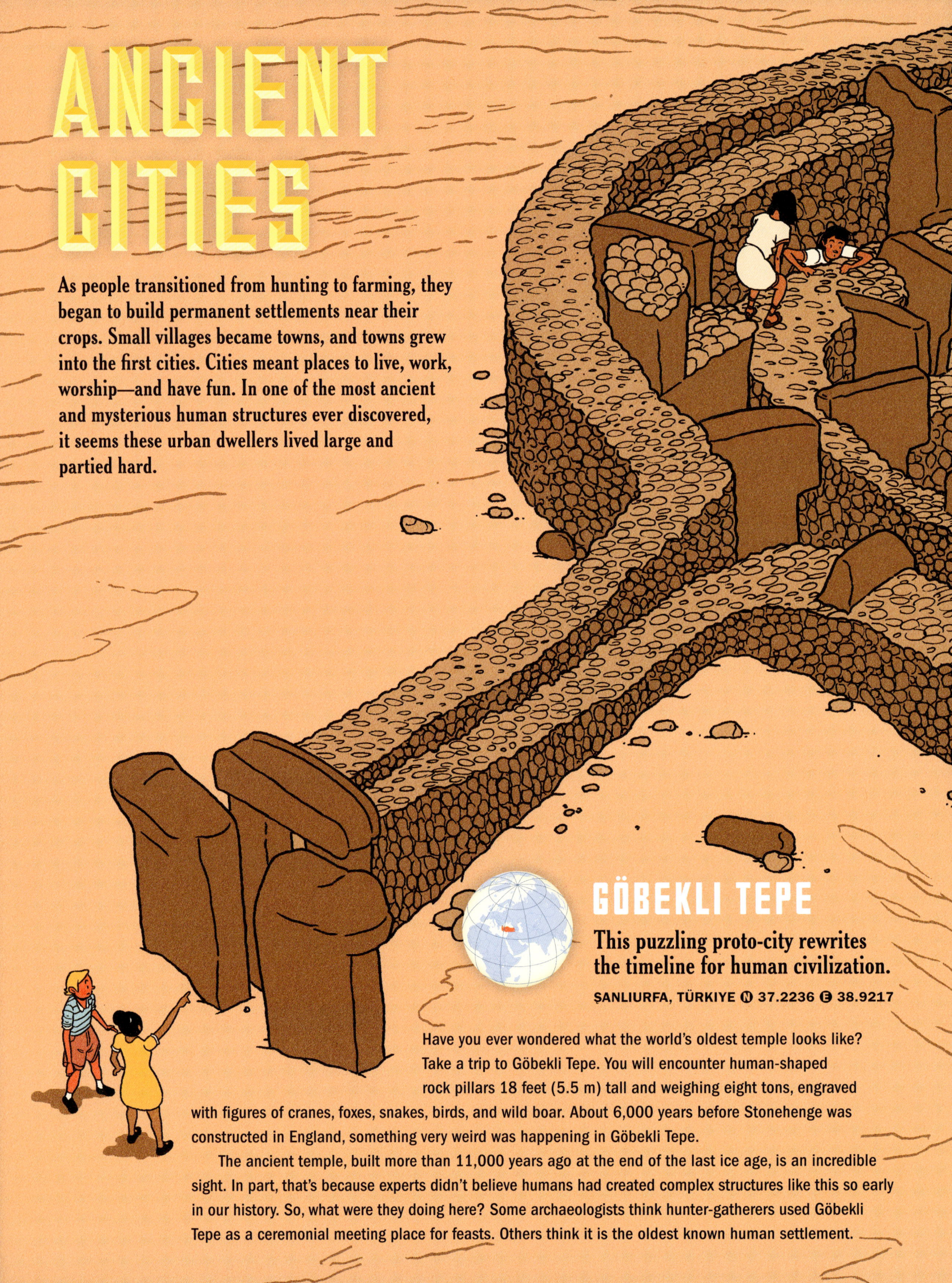

ANCIENT CITIES

As people transitioned from hunting to farming, they began to build permanent settlements near their crops. Small villages became towns, and towns grew into the first cities. Cities meant places to live, work, worship—and have fun. In one of the most ancient and mysterious human structures ever discovered, it seems these urban dwellers lived large and partied hard.

GÖBEKLI TEPE

This puzzling proto-city rewrites the timeline for human civilization.

ŞANLIURFA, TÜRKIYE Ⓝ 37.2236 Ⓔ 38.9217

Have you ever wondered what the world's oldest temple looks like? Take a trip to Göbekli Tepe. You will encounter human-shaped rock pillars 18 feet (5.5 m) tall and weighing eight tons, engraved with figures of cranes, foxes, snakes, birds, and wild boar. About 6,000 years before Stonehenge was constructed in England, something very weird was happening in Göbekli Tepe.

The ancient temple, built more than 11,000 years ago at the end of the last ice age, is an incredible sight. In part, that's because experts didn't believe humans had created complex structures like this so early in our history. So, what were they doing here? Some archaeologists think hunter-gatherers used Göbekli Tepe as a ceremonial meeting place for feasts. Others think it is the oldest known human settlement.

People at Göbekli Tepe made huge vats of porridge and beer with wheat, and the bones of animals like leopards, deer, foxes, and giant cattle called aurochs have all been found at the site. Between the building, feasting, and farming, Göbekli Tepe sure looks a lot like the world's first city. Researchers agree that the people here probably threw some epic festivals.

Excavation of the three different levels has not yet shown whether humans ever lived at the site. But the gigantic temple spans more than 12 football fields, so archaeologists are still digging for answers.

KNOW BEFORE YOU GO A wooden walkway takes visitors through the ongoing archaeological dig. Nearby, the Şanlıurfa Museum shows off some of the best objects archaeologists have uncovered.

WRITING

When people started farming and formed cities, they soon needed new ways to keep track of information: How much grain was left? How many sheep were they owed? How do we make that tasty recipe? It required writing things down. In at least four different early civilizations (Mesopotamia, Egypt, China, and Mesoamerica), new city-settlements led to the invention of writing.

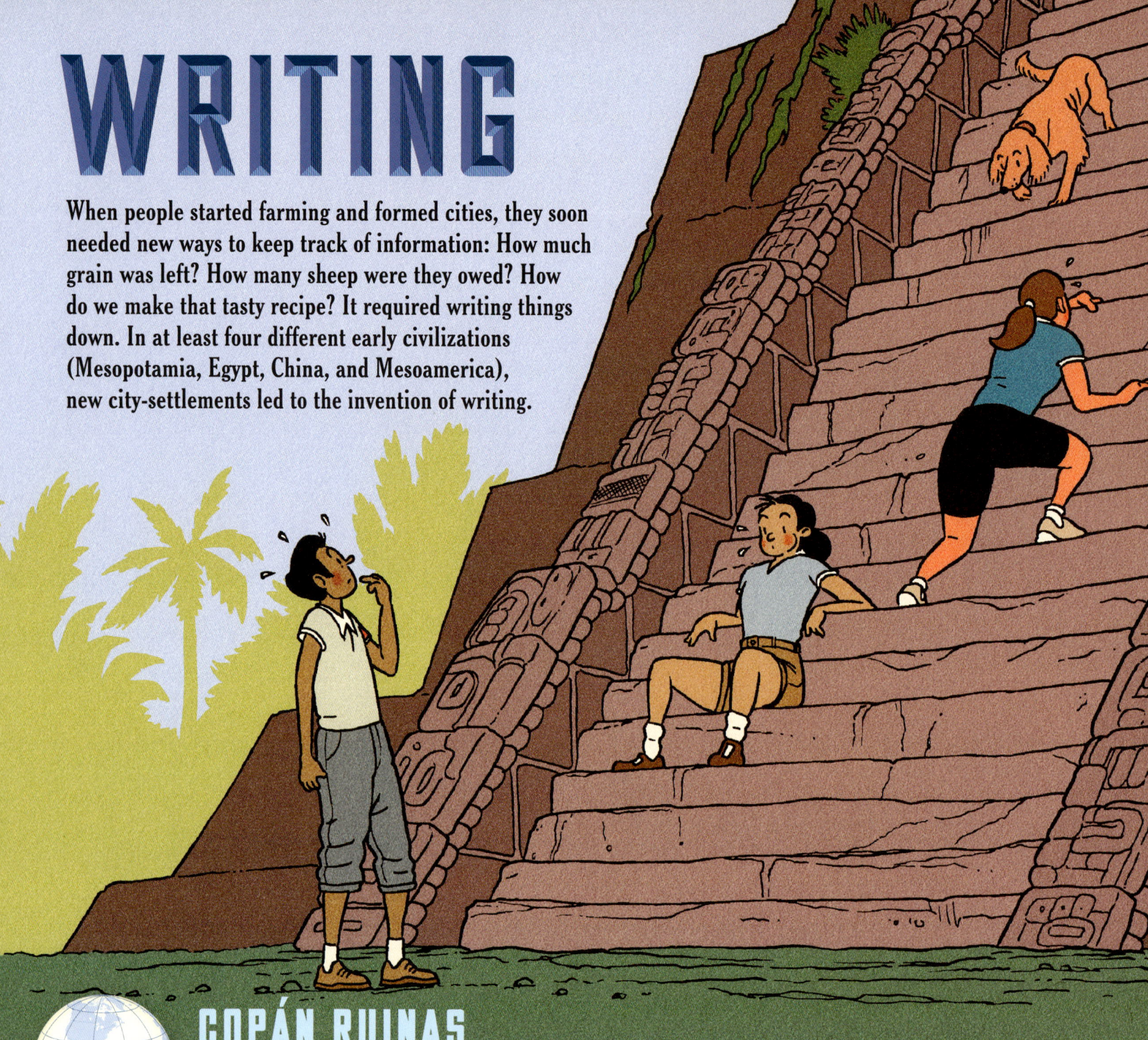

COPÁN RUINAS

A stairway of stories

COPÁN RUINAS, HONDURAS Ⓝ 14.8375 Ⓦ 89.1425

Sitting 2,188 feet (667 m) above sea level in a lush mountain valley in western Honduras is a well-preserved ancient city. Here is Copán, which 1,100 years ago was an extraordinary political and religious center of the Mayan empire. Ruins of carved stone columns and an ancient stadium provide evidence of a rich Mayan culture. But your eye will most likely be drawn to the incredible 98-foot-high (30 m) stairway of Temple 26.

Along its 63 steps are more than 1,100 pictures carved into stone. No, this isn't graffiti—it's hieroglyphics, a system of writing that uses pictures and symbols to record information. Mayan hieroglyphics haven't been used for hundreds of years, so reading them can be tricky—especially if the stone has eroded (faded away) over time.

It didn't help that earthquakes had toppled half the stones, and archaeologists needed to rearrange them while trying to decipher the meaning. Only later did they realize the hieroglyphs were written in chronological order. Oops! Eventually they discovered that the long Mayan text mostly listed the names of the ancient kings and important events during their rule.

KNOW BEFORE YOU GO There are tourist buses that take people to the ruins, but you can also hop on a local, colorfully painted "chicken bus" to get there.

ORACLE BONES

Ancient bones with a hidden message

YINXU MUSEUM, ANYANG, CHINA ⓝ 36.1227 ⓔ 114.3257

In 1899, Wang Yirong, a famous scholar of ancient writing, had soup for lunch. As the story goes, he went to take a sip, and in his bowl he discovered a piece of bone with strange inscriptions on it.

What Wang Yirong saw amazed him. The inscriptions turned out to be a form of ancient writing from thousands of years earlier. Scholars would discover that these "oracle bones" used more than 1,200 different symbols. They were prayers to change the weather, cure a toothache, or win a battle. They are some of the earliest known writing found in China. In the early 1900s, thousands of these cow and turtle bones were unearthed in the ancient city of Yinxu, capital of the Shang dynasty (1300–1046 BCE).

Why was the bone in Wang Yirong's soup? In the 3,000 years since the oracle bones were first carved, villagers had found them and recognized they were special. They saved them, ground them into dust for use as medicine, or added them to food to help cure illness. Good thing Wang decided to look down before slurping!

KNOW BEFORE YOU GO Yinxu is one of China's largest archaeological sites. Take some extra time to walk around the museum and you'll see the royal palace and tombs along with these famous bones.

"MY TOOTH HURTS! FIX IT!"

PAPER

As writing became more common, people experimented with all sorts of materials to scribble on: clay blocks, animal skins (parchment), and mats of crushed reeds (papyrus). But writers wanted something flat, light to carry, and cheap to make. If only someone could find a way to turn old junk into a material to record our ideas on . . .

HEMP

BAMBOO

MULBERRY

QILIANG VILLAGE

A 2,000-year-old paper recipe gets a new ingredient.

QILIANG VILLAGE, CHINA
Ⓝ 34.1140 Ⓔ 108.4542

You read books (like this one) printed on paper. You buy things with money made of paper. And, yes, when you go to the bathroom, you even wipe with paper.

Thank ancient Chinese inventors for all that. More than 2,000 years ago, a nobleman named Cai Lun took trash like old rags and ripped fishing nets, mixed them with bamboo, hemp, and mulberry tree bark, chopped it all up, boiled it in water, and pressed the pulp together. The result? Paper! Paper meant more ideas could be shared more easily than by writing methods like carving in wood or printing on expensive silk.

Today you can see craftspeople in the village of Qiliang making paper in a similar way. Only instead of using mulberry trees, these papermakers use panda poop. That's right, panda poop. The bamboo fibers found in the poo piles are plucked out, boiled, and then molded into flat sheets. This ancient paper is fancy stuff, costing 10 to 20 times more than normal paper. It may be made from poop, but don't wipe with it!

KNOW BEFORE YOU GO In Qiliang, you can take a class and learn to make paper just as they have for more than 2,000 years.

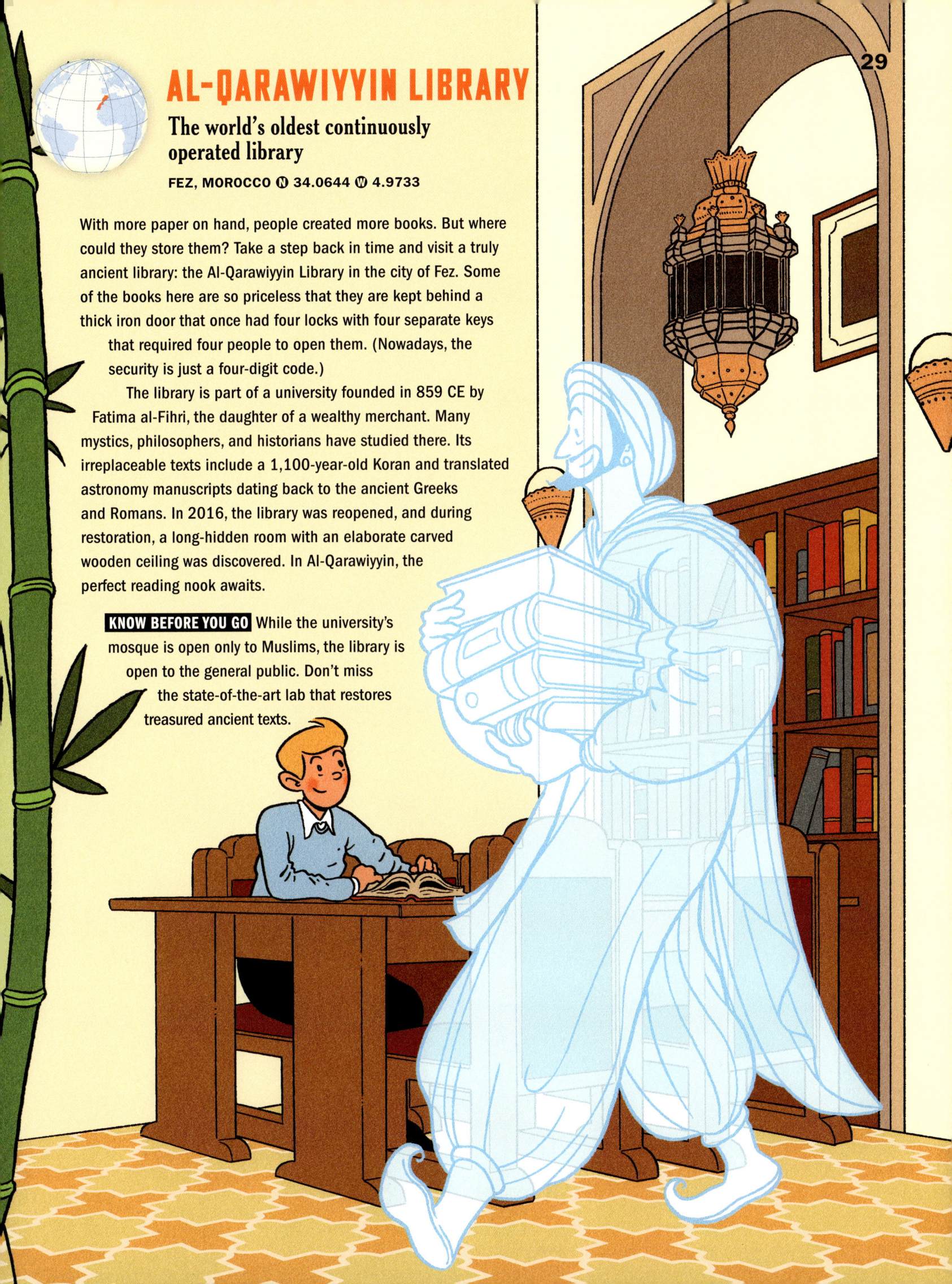

AL-QARAWIYYIN LIBRARY

The world's oldest continuously operated library

FEZ, MOROCCO Ⓝ 34.0644 Ⓦ 4.9733

With more paper on hand, people created more books. But where could they store them? Take a step back in time and visit a truly ancient library: the Al-Qarawiyyin Library in the city of Fez. Some of the books here are so priceless that they are kept behind a thick iron door that once had four locks with four separate keys that required four people to open them. (Nowadays, the security is just a four-digit code.)

The library is part of a university founded in 859 CE by Fatima al-Fihri, the daughter of a wealthy merchant. Many mystics, philosophers, and historians have studied there. Its irreplaceable texts include a 1,100-year-old Koran and translated astronomy manuscripts dating back to the ancient Greeks and Romans. In 2016, the library was reopened, and during restoration, a long-hidden room with an elaborate carved wooden ceiling was discovered. In Al-Qarawiyyin, the perfect reading nook awaits.

KNOW BEFORE YOU GO While the university's mosque is open only to Muslims, the library is open to the general public. Don't miss the state-of-the-art lab that restores treasured ancient texts.

PRINTING PRESS

Religious monks around the world created some of the earliest versions of books. When written by hand, a single book could take a year or more to make—a faster way was needed. Enter the printing press! People wanted this technology so badly that it was invented three separate times, first in China, then in Korea, and finally in Europe. Each was a step closer to printing the future.

CHEONGJU EARLY PRINTING MUSEUM

The monks' printing press

CHEONGJU, SOUTH KOREA

Ⓝ 36.6443 Ⓔ 127.4719

Get a potato. Cut it in half and carve a shape (like a star) on it. Press the star into ink and then onto paper. You just made a block print! And, by inking the star again, you can use it over and over, like a stamp. People have been making block prints for 4,500 years. In the early 1200s, Korean monks printed a full version of the Buddhist holy texts this way, only they used wood, not potatoes. After making 81,258 printing blocks with 52,330,152 individual characters (symbols), they printed 6,568 full volumes. Whew! That's a lot of blocks—and words. Next up? They were asked to make an even longer book. But how?

A monk named Choe Yun-ui had a solution. In 1250, he made metal versions of the Korean characters. These metal characters could be moved around, placed into a frame, covered in ink, and pressed onto paper to print. The characters could then be rearranged and a new page printed. "Moveable type" was a revolution in information-making, one that would be repeated in Europe two centuries later by a goldsmith whose name became even more famous worldwide than Choe Yun-ui's.

KNOW BEFORE YOU GO Visit the museum during a casting demonstration. It's really cool—well, hot—because it involves pouring molten metal. And don't mind the creepy monk mannequins everywhere!

GUTENBERG PRESSES

Pressing "Print" on an information revolution

MUSEUM PLANTIN-MORETUS, ANTWERP, BELGIUM ⓝ 51.2183 ⓔ 4.3978

The first thing you notice as you walk into the Museum Plantin-Moretus is the faint smell of ink and wood. In this former workshop is a replica of an 8-foot-tall (2.4 m) printing press created by a goldsmith named Johannes Gutenberg. The Gutenberg press is a 570-year-old technology that transformed its time as much as the internet has changed ours.

Gutenberg's metalwork skills helped him cast the little metal letters for his press. Germany was a European center of mining, so metal was abundant. The language he printed helped him innovate this technology in another way, too. Early printing presses in China and Korea required thousands of individual characters, but Gutenberg printed the Bible and other texts in Latin. He needed just 290 individual characters for that language, so the Gutenberg press could print books and pamphlets much more easily.

Before the printing press, there were only tens of thousands of books in Europe. Fifty years after Gutenberg's invention, an estimated nine million books were in print. Books were suddenly everywhere!

KNOW BEFORE YOU GO This is the only museum on the UNESCO World Heritage list. Besides the printing press, there is a seventeenth-century library, a bookshop, and a beautiful garden for you to explore.

ASTRONOMY

The printing press sparked revolutions in both religion and science. Ancient texts once locked away in distant libraries could now be reprinted and shared. This included books with mathematical data about the cosmos: Eighteen years after Gutenberg printed the Bible, the first astronomy book was printed. The printing press thus helped create a renaissance (rebirth) of astronomical discovery!

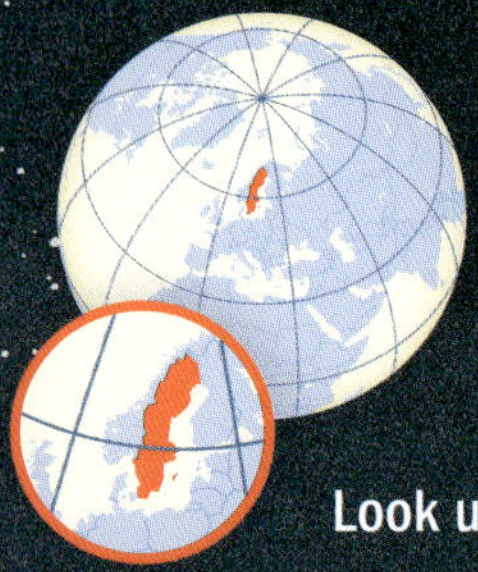

STJERNEBORG

The island observatory where a metal-nosed man mapped the stars

VEN ISLAND, SWEDEN Ⓝ 55.9070 Ⓔ 12.6970

Look up into the night sky. Could you imagine making a map of all those stars? It would take a *really* long time. Just ask Tycho Brahe, a Danish astronomer who spent more than 20 years in his castle observatory recording a thousand stars. He used astronomical instruments like a giant wooden sphere to record the stars' locations. (Telescopes hadn't been invented yet!)

Tycho was a tough customer. His nose was cut off in a sword fight with his cousin over who was better at mathematics! He wore a nose made of metal the rest of his life. Later, in 1577, Tycho noticed a light streaking across the evening sky (a comet) while returning from a day of fishing. He would theorize correctly that this comet moved in a giant loop around the sun, something that even Galileo hadn't thought of yet.

At Stjerneborg castle, you can explore part of the original observatory where Tycho made his revolutionary findings that inspired famous astronomers such as Johannes Kepler and Sir Isaac Newton.

KNOW BEFORE YOU GO Ven is a short boat ride from Copenhagen, and you can explore the whole island by bike! Be sure to play games in the castle gardens and visit the ruined paper mill where Tycho made his own books.

DANIEL K. INOUYE SOLAR TELESCOPE

Watch the sun explode from the top of the largest dormant volcano in the world.

HALEAKALĀ, MAUI, HAWAI'I Ⓝ 20.7068 Ⓦ 156.2562

Sitting atop a massive 10,062-foot-tall (3,067 m) volcano on the island of Maui is what looks like a giant white ice cream cone. Of course, this is no dessert. Instead, it's the largest solar telescope in the world.

Just like how Tycho Brahe mapped the stars, the Daniel K. Inouye Solar Telescope (DKIST) maps the sun. Inside the DKIST's domed top is a 13-foot-long (3.96 m) primary mirror that peers into the sun's depths. It captures images of enormous solar explosions, known as coronal mass ejections, that shoot out clouds of hot plasma (energized gas) and bits of magnetic fields into space. Just one such explosion has the power of a billion atomic bombs! When these solar ejections reach Earth, they sometimes create radio, magnetic, and thermal disruptions that can destroy satellites and power grids, effectively turning off our lights.

Astronomers hope to use the DKIST to understand how and why these eruptions happen. But learning the secrets of the sun may do more than just protect us from solar explosions—one day it may help us power the planet.

KNOW BEFORE YOU GO While you can't tour the telescope itself yet, you can drive up a steep and winding road to the top of Haleakalā volcano and enjoy beautiful views of both the telescope and the island.

FUSION

People had been observing the stars for millennia, but what powered stars remained mysterious. In 1938, a young physicist named Hans Bethe looked to our own star, the sun, to prove that it is a gigantic nuclear reactor. He theorized that a process called fusion (smushing atoms together) is what produces all that energy. Scientists hope to re-create fusion power here on Earth, using star energy to power the planet.

CRYOSTAT CHAMBER— COLDER THAN PLUTO!

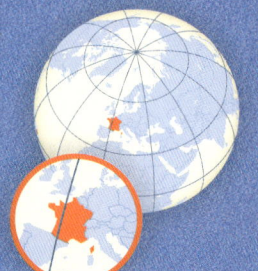

INTERNATIONAL THERMONUCLEAR EXPERIMENTAL REACTOR

A magnetic donut filled with sun jelly

SAINT PAUL-LEZ-DURANCE, FRANCE Ⓝ 43.7083 Ⓔ 5.7774

Sitting outside on a sunny day makes you hot. That's because the sun is a huge explosive sphere so powerful that it warms and lights our planet from 93 million miles (149.7 million km) away. With a temperature of more than 27 million°F (15 million°C), that distant ball of glowing plasma is what keeps us alive and toasty.

What if you could make a tiny sun here on Earth? Think of the energy that would become available! It could power your house, your city, and perhaps even an entire country. At the International Thermonuclear Experimental Reactor (ITER), scientists are hoping to do just that. ITER is trying to produce energy using nuclear fusion, a process in which two atoms are squished into one new atom, releasing enormous energy in the process. But how do you control a tiny sun? What you need is a gigantic magnetic donut.

Called a tokamak, this machine will produce such strong magnetic fields that superhot plasma will float in its center like 180 million°F (100 million°C) jelly inside a donut. The heat energy produced could be turned into electricity. That's the idea, anyway. Scientists are still working on it, and some think it's an impossible task. But if they succeed, these massive magnets could one day power our whole world.

FISSION VS FUSION

You may know that some countries use nuclear power plants for energy. That type of nuclear energy is from fission.

Fission = taking one large atom and splitting it into two atoms. This process releases a lot of energy. That energy is converted to electrical energy that is used to power homes, buildings, and more.

Fusion = taking two or more atoms and fusing them together to create one new atom. This process requires extremely high temperatures and pressures, but when it works, it releases HUGE amounts of energy. Fusion is the ultimate in clean energy!

KNOW BEFORE YOU GO Once a year on Open Doors Day in April, ITER welcomes hundreds of visitors. Kids over the age of ten can explore the construction site along with their parents.

RADIOACTIVE BLANKET
KEEPS NEUTRONS IN

THE 8,000-TON
STEEL DONUT

"DOUBLE PANCAKES"
OF MORE MAGNETS

CENTRAL SOLENOID
MAGNET TO HOLD
STAR JELLY

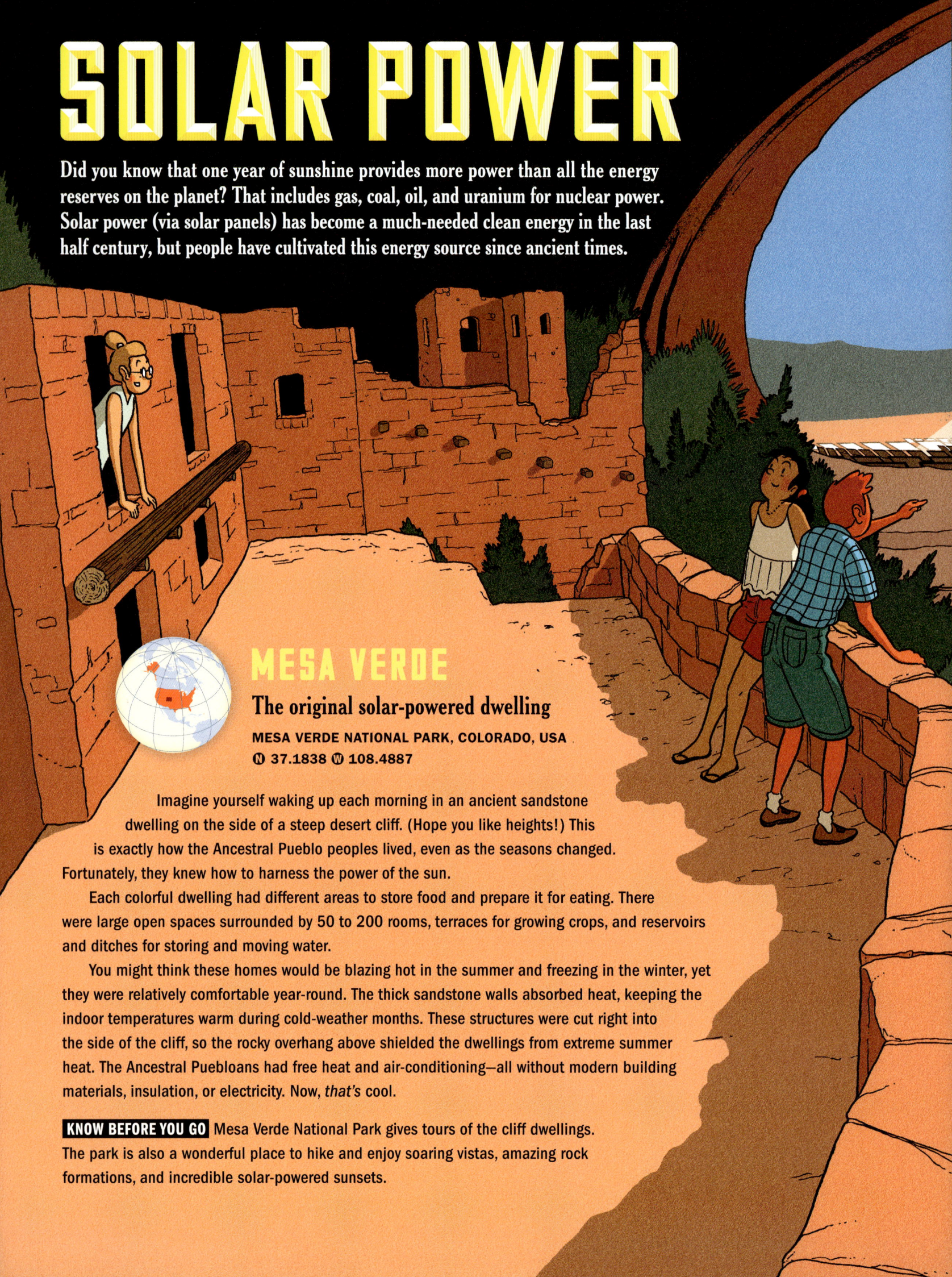

SOLAR POWER

Did you know that one year of sunshine provides more power than all the energy reserves on the planet? That includes gas, coal, oil, and uranium for nuclear power. Solar power (via solar panels) has become a much-needed clean energy in the last half century, but people have cultivated this energy source since ancient times.

MESA VERDE

The original solar-powered dwelling

MESA VERDE NATIONAL PARK, COLORADO, USA
Ⓝ 37.1838 Ⓦ 108.4887

Imagine yourself waking up each morning in an ancient sandstone dwelling on the side of a steep desert cliff. (Hope you like heights!) This is exactly how the Ancestral Pueblo peoples lived, even as the seasons changed. Fortunately, they knew how to harness the power of the sun.

Each colorful dwelling had different areas to store food and prepare it for eating. There were large open spaces surrounded by 50 to 200 rooms, terraces for growing crops, and reservoirs and ditches for storing and moving water.

You might think these homes would be blazing hot in the summer and freezing in the winter, yet they were relatively comfortable year-round. The thick sandstone walls absorbed heat, keeping the indoor temperatures warm during cold-weather months. These structures were cut right into the side of the cliff, so the rocky overhang above shielded the dwellings from extreme summer heat. The Ancestral Puebloans had free heat and air-conditioning—all without modern building materials, insulation, or electricity. Now, *that's* cool.

KNOW BEFORE YOU GO Mesa Verde National Park gives tours of the cliff dwellings. The park is also a wonderful place to hike and enjoy soaring vistas, amazing rock formations, and incredible solar-powered sunsets.

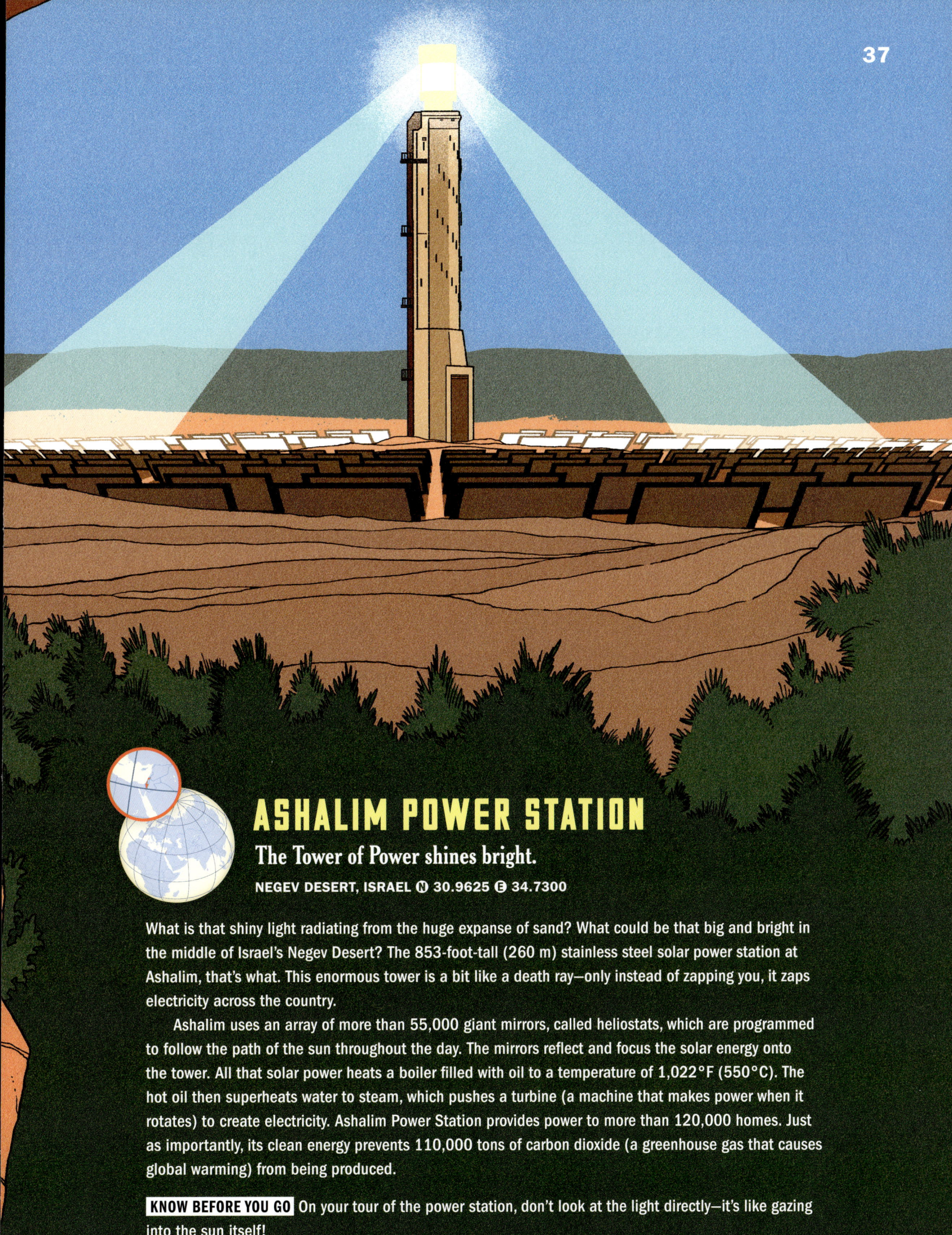

ASHALIM POWER STATION

The Tower of Power shines bright.

NEGEV DESERT, ISRAEL Ⓝ 30.9625 Ⓔ 34.7300

What is that shiny light radiating from the huge expanse of sand? What could be that big and bright in the middle of Israel's Negev Desert? The 853-foot-tall (260 m) stainless steel solar power station at Ashalim, that's what. This enormous tower is a bit like a death ray—only instead of zapping you, it zaps electricity across the country.

Ashalim uses an array of more than 55,000 giant mirrors, called heliostats, which are programmed to follow the path of the sun throughout the day. The mirrors reflect and focus the solar energy onto the tower. All that solar power heats a boiler filled with oil to a temperature of 1,022°F (550°C). The hot oil then superheats water to steam, which pushes a turbine (a machine that makes power when it rotates) to create electricity. Ashalim Power Station provides power to more than 120,000 homes. Just as importantly, its clean energy prevents 110,000 tons of carbon dioxide (a greenhouse gas that causes global warming) from being produced.

KNOW BEFORE YOU GO On your tour of the power station, don't look at the light directly—it's like gazing into the sun itself!

ANCIENT RENEWABLES

Though the Ancestral Puebloans made good use of solar power at Mesa Verde, ancient civilizations didn't get power from the sun alone. Long before the First Industrial Revolution (1760–1840), almost all power came from renewables (power sources that are replenished over time). Wind, water, and animals were a few of the ways the ancient world made things move.

NASHTIFAN WINDMILLS

Vertical windmills that still work a thousand years later

NASHTIFAN, IRAN Ⓝ **34.4316** Ⓔ **60.1748**

Soaring 65 feet (19.8 m) above the desert, 24 mud-covered adobe walls don't just protect the people of Nashtifan from the gale-force winds during monsoon season. They also funnel the winds to generate energy. In this small village in northeastern Iran, renewable energy has been in use for ten centuries!

These windmills, some of the first ones ever built, are unique in their construction. Unlike the iconic four-blade towers you might find in the Netherlands (or on a mini-golf course), the ancient Nashtifan windmills are vertical, with long wooden slats that catch the wind.

Here's how they work: The wind flows into the large clay structure and moves the vertical wooden vanes. This rotates a grindstone at the bottom, crushing wheat into flour used to bake bread. A full day of wind can produce 165 pounds (74.8 kg) of flour. We can certainly "toast" to that!

KNOW BEFORE YOU GO Ask if Mohammad Etebari is around—he's the caretaker. Try to go on a windy day so you can see the windmills in action.

NORIAS OF HAMA

Enormous waterwheels of great power

HAMA, SYRIA Ⓝ **35.1353** Ⓔ **36.7531**

Standing next to the massive norias (waterwheels) of Hama, you might feel a bit small. Some of these gigantic structures are 69 feet (21 m) tall . . . as big as a brontosaurus is long! Today, when you need water, you turn on the faucet. But many centuries ago, people didn't have that luxury. They had to get their water by pulling it directly out of a river or other freshwater source. Think how many buckets it would take to fill up a bathtub!

Norias were invented around the fifth century CE to solve this problem. The norias here in the Orontes River in Syria could raise 660 gallons (2,498 L)—or 16 bathtubs full—in just one minute. As the force of the river's current turned the wheel, its boxlike compartments filled up with water and poured into a nearby aqueduct (a channel for water). From there, the water flowed to the village or out to the fields to irrigate crops.

The waterwheels still turn today and make enormous groaning noises. Not only are they as big as a dinosaur, but they sound like one, too.

KNOW BEFORE YOU GO Due to war in Syria, the waterwheels have been under threat, but as of 2020 there has been a concerted effort by locals to preserve and protect them.

PUMPS

Every human is powered by the pump inside them (their heart), and our whole world is powered by pumps, too. It might not sound exciting, but the pump is one of the most important inventions of all time. Pumps move water for farms, gas in combustion engines, and coolant in refrigerators. Figuring out better ways to transport water and other fluids opened vast new possibilities for innovation.

HYPERION WATER RECLAMATION PLANT

This ancient Greek technology "screws up."

LOS ANGELES, CALIFORNIA, USA Ⓝ 33.9250 Ⓦ 118.4297

Ever wonder what happens when you flush the toilet? Head to the Hyperion Water Reclamation Plant in Los Angeles, California, and find out.

Standing among the huge globe-shaped tanks storing methane gas, you'll immediately notice the miles of pipe overhead—and the faint odor of poop. But you'll also learn one of the more ingenious ways to move water, or in this case, raw sewage. Based on a 2,000-year-old invention by ancient Greek scientist Archimedes of Syracuse, this pump moves wastewater from low to high areas. It's shaped like a giant screw, and as it turns, water is trapped in each blade of the screw and lifted up and up until it overflows out of the top.

The Hyperion Water Reclamation Plant provides clean water for many uses, like irrigation, industry, and more. But we don't particularly recommend drinking it!

KNOW BEFORE YOU GO You can take a tram tour of the wastewater plant, but beware: It's not the nicest-smelling tour.

IJMUIDEN PUMPING STATION

Pumps that keep out the sea

IJMUIDEN, NETHERLANDS Ⓝ **52.4673** Ⓔ **4.6058**

When you dig in the sand at the beach, ever notice that the deeper you go, the more water flows in? Water is difficult to contain, and the Dutch know this fact all too well. More than 25 percent of their country is below sea level. To keep the water off their land (and out of their houses), they rely on pumps. *Really big* pumps.

One of the largest pumps in the world is located at IJmuiden Pumping Station, which connects the capital city of Amsterdam to the North Sea Canal. Originally, the Dutch pumped water out of the lowlands with windmill-powered Archimedes screw pumps. Today, the IJmuiden station uses six massive electric pumps that transport more than 950,000 gallons per minute (3,596,141 L/minute). That's like pumping out more than an Olympic-sized swimming pool every single minute. (These machines could fill your backyard pool in one and a half seconds.) The three-story, 120-ton pumps prevent the land from flooding, while letting boats float on canals into the port of Amsterdam.

KNOW BEFORE YOU GO The IJmuiden station is not normally open for public tours, but you can visit the nearby Wouda pump, the largest steam-pumping station ever built!

DISCHARGE

SHAFT

IMPELLER

BEARING ★

★ NOT AN EXACT DIAGRAM OF IJMUIDEN PUMP

STEAM ENGINES

Transporting water required pumps, but water could also help power those pumps. In the late 1690s, engineer Thomas Savery did just that. By burning wood or coal, Savery heated water into steam, creating a vacuum that pumped water out of the ground. Savery's steam pump led to an even more powerful concept: the steam engine.

SMETHWICK ENGINE

The oldest working steam engine in the world

THINKTANK BIRMINGHAM SCIENCE MUSEUM, BIRMINGHAM, UK Ⓝ 52.4829 Ⓦ 1.8861

Before trains crisscrossed England, canals moved goods across the country. To move ships through these canals, special areas were needed where water could be raised and lowered. Known as locks, they floated ships up and down to allow them to navigate waterways that flowed at different levels. The Smethwick Engine was designed to power a canal's locks.

Built in 1779, the engine was the size of a three-story building and could pump more than 1,500 buckets of water into the locks each minute. This allowed 250 boats a day to pass through a canal from the city of Wolverhampton to the city of Birmingham. It was one of the first major uses of a steam engine in England.

So how did it work? Steam expands rapidly, creating pressure. When you direct that pressure at a piston (a cylinder inside a shaft), the piston is shoved upward, which moves a wheel and pumps water into a pipe. When that steam cools, pressure decreases, creating a vacuum and sucking the piston back down. Repeating this process over and over creates a pumping action that lifts water.

KNOW BEFORE YOU GO This 240-year-old steam engine is still puffing away at its current location at the Thinktank Birmingham Science Museum. Don't forget to also check out the Science Garden. You can stand under a see-through globe and experience rain without getting wet!

PS SKIBLADNER

The world's oldest paddle steamer

LAKE MJØSA, NORWAY ⓝ 60.7984 ⓔ 10.6969

Sitting majestically like a beautiful swan on a lake is a sleek, white-hulled ship: the PS *Skibladner*. As you climb aboard, you'll notice the polished wooden planks have been worn by more than 160 years of service. The big funnel in the middle of the deck pumps out big clouds of steam coming from the engine below.

The way the ship puffs along, you wouldn't know that you're on the world's oldest paddle steamer. To make it move, the ship's steam engine pushes a piston, which helps turn two giant paddle wheels. The paddle wheels push water behind the ship, propelling it forward at a steady speed of 12 knots (about 14 mph or 22 kph). The ship was revolutionary in its time, providing safe and fast movement for people and goods.

KNOW BEFORE YOU GO You can still take a ride on the PS *Skibladner*. Visit in spring or summer—during winter Lake Mjøsa is covered in ice.

TRAINS

As steam engines became more efficient, people looked for new ways to use them. Add a steam engine to a big box with wheels, and you can transport passengers and heavy cargo across land—provided you already have the tracks in place.

PUFFING BILLY

The world's oldest steam locomotive

SCIENCE MUSEUM, LONDON, UK

Ⓝ 51.4975 Ⓦ 0.1747

CLACK-clack, CLACK-clack goes the huge mechanical contraption as it rolls slowly along a metal track. Steam puffs out of a tall pipe, and long, vertical metal arms attached to the wheels pump up and down. Ever so gradually, the locomotive chugs forward on its eight huge wheels. The people watching nearby cheer as it passes.

It is the 1800s, and you need to move a huge load of coal from a mine to the port where ships are docked. A couple thousand pounds is way too heavy to carry by hand—and that's where trains come in. By using a steam engine to turn the wheels of their vehicle, engineer William "Billy" Hedley and his partners built one of the first functional trains in 1813. They called it Puffing Billy.

It wasn't the most elegant design: The train was made of cast iron and was so heavy it broke its metal rails. It was awkward, dangerous, and traveled at only 5 mph (8 kph)—the speed of a brisk walk. Even so, it was still a whole lot better than carrying all that coal by yourself!

KNOW BEFORE YOU GO Puffing Billy is on display in London's Science Museum. Just don't get accidentally detoured to Australia to see a train of the same name!

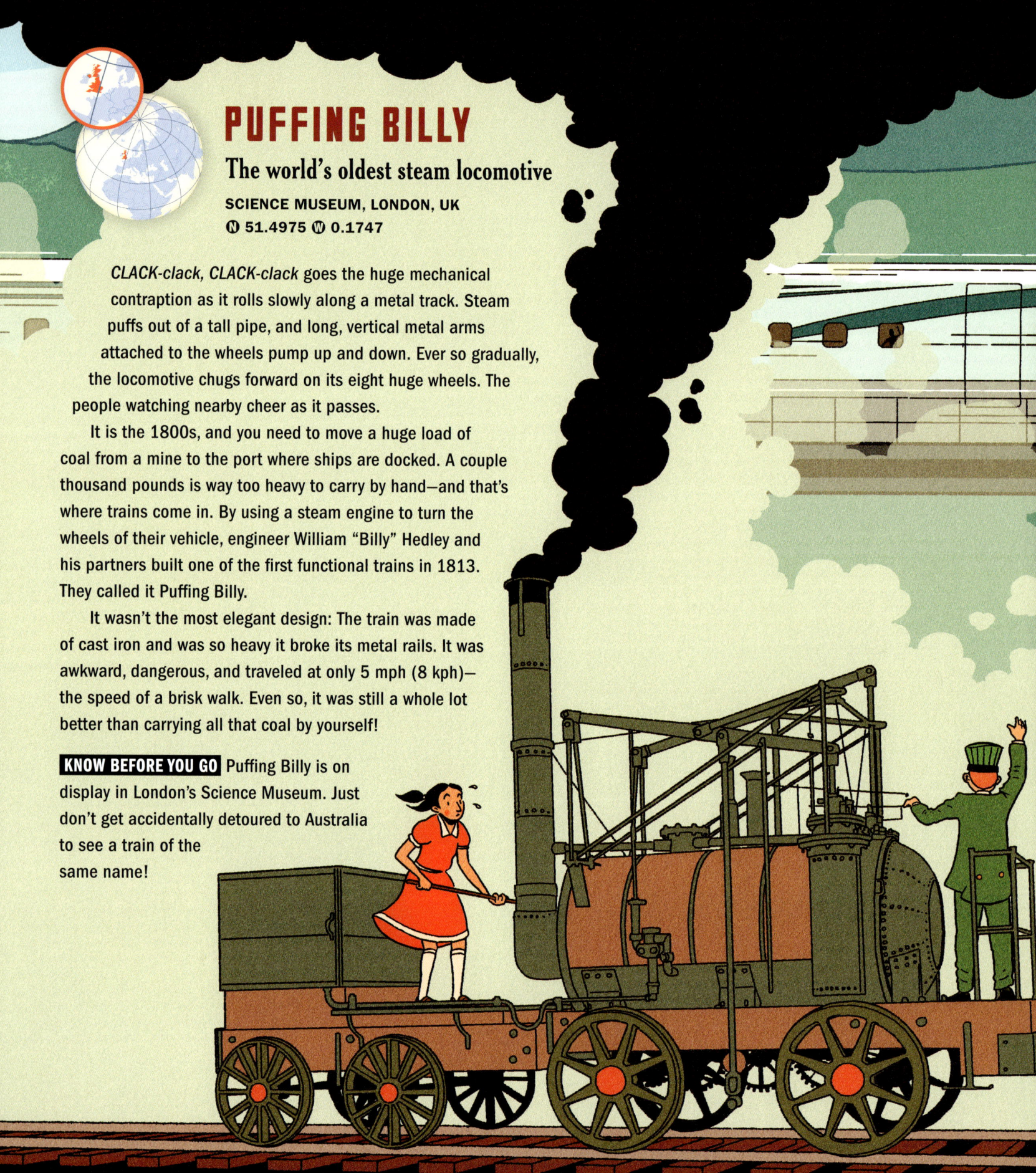

MAGNETS

TRAIN HOVERS
4 INCHES OFF
THE GROUND!

JAPAN'S FLOATING TRAINS

These super-speedy trains are held up by magnets.

SHINAGAWA STATION, TOKYO, JAPAN Ⓝ 35.6286 Ⓔ 139.7392

Whoosh! Did you see that long blue-and-white blur? It was so fast . . . could it really have been a train? Yes, Japan's superconducting maglev trains go a really, *really* fast 375 mph (603.5 kph). How is this possible? The Japanese have been researching magnetic levitation (known as maglev) technology since the 1970s. Maglev trains don't use rails, and their rubber wheels touch the ground only when they're starting or stopping. Instead, they hover almost 4 inches (10.2 cm) over their track and travel safely within a concrete guideway. The secret is magnetism. When you hold two magnets in your hand, they either attract or repel each other. Magnetic levitation works the same way. The train has powerful superconducting magnets attached to the bottom of each car. Magnetic coils are built into the track's concrete guide walls for balance, levitation, and propulsion. Energy is supplied to the coils in the walls, which causes the train's magnets to alternately repel and attract. This allows the train to speed up, slow down, or come to a stop. It's about the closest you can get to being on a hoverboard.

KNOW BEFORE YOU GO Japan is still building the 177-mile-long (284.9 km) Chūō Shinkansen maglev line between Tokyo and Nagoya. In the meantime, there's a mini maglev train called the Linimo Line that runs for 5.5 miles (8.9 km) at 60 mph (96.6 kph).

MINES

Trains kicked the First Industrial Revolution into gear. But what powered these locomotives and other engines? Coal dug up from mines. Steam engines pumped water out of the mines and steam trains transported the minerals. Each technology reinforced the other. Mines still provide our world with many resources: ores for metals, lithium for batteries, and much more.

PIONEER TUNNEL

PIONEER TUNNEL COAL MINE

A mine car ride into the past

ASHLAND, PENNSYLVANIA, USA Ⓝ 40.7780 Ⓦ 76.3522

Put on your hard hat, hop into the mine car, and hang on. You're about to take a ride back in time.

As you descend 400 feet (121.9 m) underground and 1,800 feet (548.6 m) into the side of Mahanoy Mountain, it gets dark and a bit cold. Shovel marks cover the sides of the tunnel, and if you close your eyes, you can imagine how miners used to dig coal. They spent long hours every day chipping away at the edges of the tunnel. Workers loaded coal into the same type of mine car, or buggy, that is used today to bring visitors into the mine. The car transported the heavy coal outside, where it was loaded onto trains.

This mine shut down in 1931, and today coal powers the US less than nuclear or renewable energy. There is a very big downside to coal: Burning it releases carbon dioxide and other pollutants, which cause climate change and damage our environment. Fortunately for our planet's health, these days only tour groups—not coal—move in and out of this mine.

KNOW BEFORE YOU GO It's dark and cold down in the mine, so bring a sweater! One of the best parts of the tour is when they turn out the lights so you can experience total darkness.

BAGGER 288 EXCAVATOR

The biggest, baddest rock mover of them all

TAGEBAU HAMBACH MINE, GERMANY Ⓝ 50.9108 Ⓔ 6.5028

You've been given a job. You have to mine rocks, soil, and sand out of an area the size of a soccer field. That would take a *lot* of people—and a *lot* of time. Or, you could just borrow one of the world's largest vehicles, the Bagger 288, and dig it all out in one day.

Longer than the Washington Monument is tall and heavier than the Eiffel Tower, this 31-story-tall and nearly 30-million-pound (13.6 million kg) machine can remove 8.5 million cubic feet (240,000 m³) of earth per day. That makes it extremely useful to mine materials located beneath ground. It picks up the rock and dirt using a 148-foot (45.1 m) steel shaft supporting a 7-story-tall bucket wheel—literally a wheel of rotating buckets. It's like a Ferris wheel, except each seat is a giant dirt scoop.

Of course, a 13,000-ton excavator doesn't travel light, and it needs reinforced roads just to move. But its massive 8,600 square feet (799 m²) of tread can take the Bagger where it needs to go . . . at just 0.5 mph (0.8 kph). Walking is about six times that fast! Giants just aren't made for speed.

KNOW BEFORE YOU GO To see the Bagger 288 in action, head to the observation point over the Hambach surface mine. Bring binoculars—but it's hard to miss.

GUNPOWDER

A thousand years ago in China, miners made an explosive discovery: Combine saltpeter (salty rock) and sulfur (rocks that smell like rotten eggs) with charcoal (burnt wood) and—BANG—you have gunpowder. Before becoming an agent of death, gunpowder was thought to be an elixir of life. And besides its use in weapons, gunpowder has another more positive use: for celebrating!

FIREWORKS TOWN

A festive city with an explosive secret

LIUYANG, CHINA Ⓝ 28.1637 Ⓔ 113.6433

Boom! Red stars shoot through the sky. *Bam!* Yellow lights flash and disappear. *Whizz!* Bright blue sparks spread out like a fan. Who doesn't love a fireworks display?

Gunpowder fireworks were created around a thousand years ago in China, and the country continues to create some of the world's largest, most colorful explosions. In the city of Liuyang, you'll find more than a thousand factories that specialize in making all kinds of pyrotechnics. It's dangerous work, and there have been deadly accidental blowups. That's because at the heart of a firework is an explosion.

Some plants in the city make the black gunpowder by mixing chemicals like potassium nitrate, sulfur, and charcoal. Other factories make the fuses. There's even a whole plant dedicated to making explosive fireworks that look like birthday cakes. (Don't eat them!) These handmade fireworks range in size from 2 inches (5 cm) to almost 4 feet (1.22 m) long, and all are launchable out of a reusable tube. *Fizz! Pop! BOOM!*

KNOW BEFORE YOU GO Liuyang has a weekly fireworks show that puts the Fourth of July to shame. Even the restaurants in Liuyang are fireworks themed.

BARA GAZI CANNON

Twenty-nine feet (8.8 m) of "Leave me alone!"

GULBARGA FORT, KALABURAGI, INDIA Ⓝ 17.3405 Ⓔ 76.8311

As you approach this imposing fourteenth-century CE fort in Kalaburagi, India, you might first notice its large domed roofs. But then you'll quickly spot something else: the nearby Bara Gazi Toph (cannon). At about 29 feet (8.8 m) long, it's tough to miss.

The almost 700-year-old cannon was built by the ruler Sultan Alauddin Hasan Bahman Shah between 1327 and 1358. Made of a metal alloy called panchdhatu (a mixture of gold, iron, zinc, silver, and copper) and using gunpowder as its blasting agent, it weighs more than 15,000 pounds (6,800 kg). Just aiming the cannon takes 20 people.

Why so big? The longer the cannon, the farther a cannonball flies. The Bara Gazi's 30-pound (13.6 kg) cannonballs could travel more than 30 miles (50 km)! While it's not clear whether this giant cannon was ever fired, just knowing it was there was enough to keep invaders away. Gulbarga Fort, which houses the cannon, was so safe that the sultan was able to found a dynasty and make Kalaburagi the capital of his vast empire in central India.

KNOW BEFORE YOU GO Wear your walking shoes. The best views from the fortress are at the top, and it's a bit of a climb up old steps. (And no, you can't fire the cannon.)

INTERNAL COMBUSTION ENGINES

In the seventeenth century, a Dutch inventor named Christiaan Huygens was inspired by cannons to design an engine powered by gunpowder. Huygens's "explosion engine" was one of the first internal combustion engines ever imagined. Today's more complex gas-powered versions move cars, planes, and ships. But combustion engines contribute to climate change, and they may one day become relics of the past.

WÄRTSILÄ-SULZER RTA96-C

An engine bigger than a house

***EMMA MÆRSK* CARGO SHIP, HOMEPORT IN DENMARK Ⓝ 55.4686 Ⓔ 10.5386**

Standing on the pier next to the massive *Emma Mærsk*, a container ship based in Denmark, probably makes you feel small. Longer than the Empire State Building is tall, and weighing more than the Statue of Liberty at nearly 377 million pounds (171 million kg), the ship might also make you wonder: What do you use to *move* this thing? If you guessed the world's biggest engine, you'd be right. At 44 feet (13.4 m) tall, the diesel fuel–powered Wärtsilä-Sulzer RTA96-C isn't built for speed—it's built to move giants.

Inside each of its 14 cylinders is a 300-ton crankshaft with a 12,000-pound (5,443 kg) piston. When the engine needs repairs, engineers climb *inside* it! The Wärtsilä-Sulzer RTA96-C propels the gigantic cargo ship at a cruising speed of 27 knots (about 31 mph or 50 kph). That may not sound so fast, but when you are moving more than 220,000 tons of cargo, it's pretty remarkable.

The *Emma Mærsk* and other container ships like her collectively contribute about 2–3 percent of the carbon dioxide emissions that are speeding up global warming. The race is on to capture the carbon emissions on ships like this or replace the oil with a cleaner-burning fuel.

KNOW BEFORE YOU GO You can track the *Emma Mærsk* online in real time! If it comes to a port near you, try to catch a glimpse.

44 FEET TALL

5 FEET TALL

MAERSK LINE

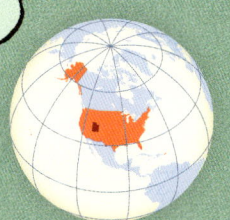

THE SPEED DEMON

They don't call it the Speed Demon for nothing.

BONNEVILLE SPEEDWAY, WENDOVER, UTAH, USA Ⓝ 40.7625 Ⓦ 113.8956

As you zoom across the hard-packed desert ground of the Bonneville Salt Flats, a cloud of dust billows behind you. You grip the wheel tight, because one wrong move could mean disaster.

You're driving the long, sleek rocket-shaped car known as the Speed Demon. It's the fastest piston-driven internal combustion engine in the world, reaching speeds of more than 470 mph (756 kph). That's more than half the speed of sound! Within the engine are cylinders where gasoline combusts (burns), creating a force that pushes metal pistons. The movement of the pistons turns the crankshaft, sending energy to the wheels.

The Speed Demon has only two pistons, but they provide *a lot* of energy. This energy is perfect for breaking the land-speed record on the wide-open land of the Utah salt flats. The car's jetlike shape is streamlined, which means it reduces friction, or drag, from the wind. With its cool winged tail, riding in the Speed Demon feels a lot like "flying" on land.

KNOW BEFORE YOU GO The Speed Demon is at the Bonneville Flats only when it is racing. Keep an eye out for the next attempt to break 500 mph (804.7 kph).

AIRPLANES

The internal combustion engine first conquered the ground, and then the sea, but these machines would soon take humans to new heights. In 1903, a mechanic named Charles E. Taylor built an aluminum combustion engine that allowed the Wright brothers to take flight. The historic trip lasted only 12 seconds, but it was the beginning of our reach toward the stars.

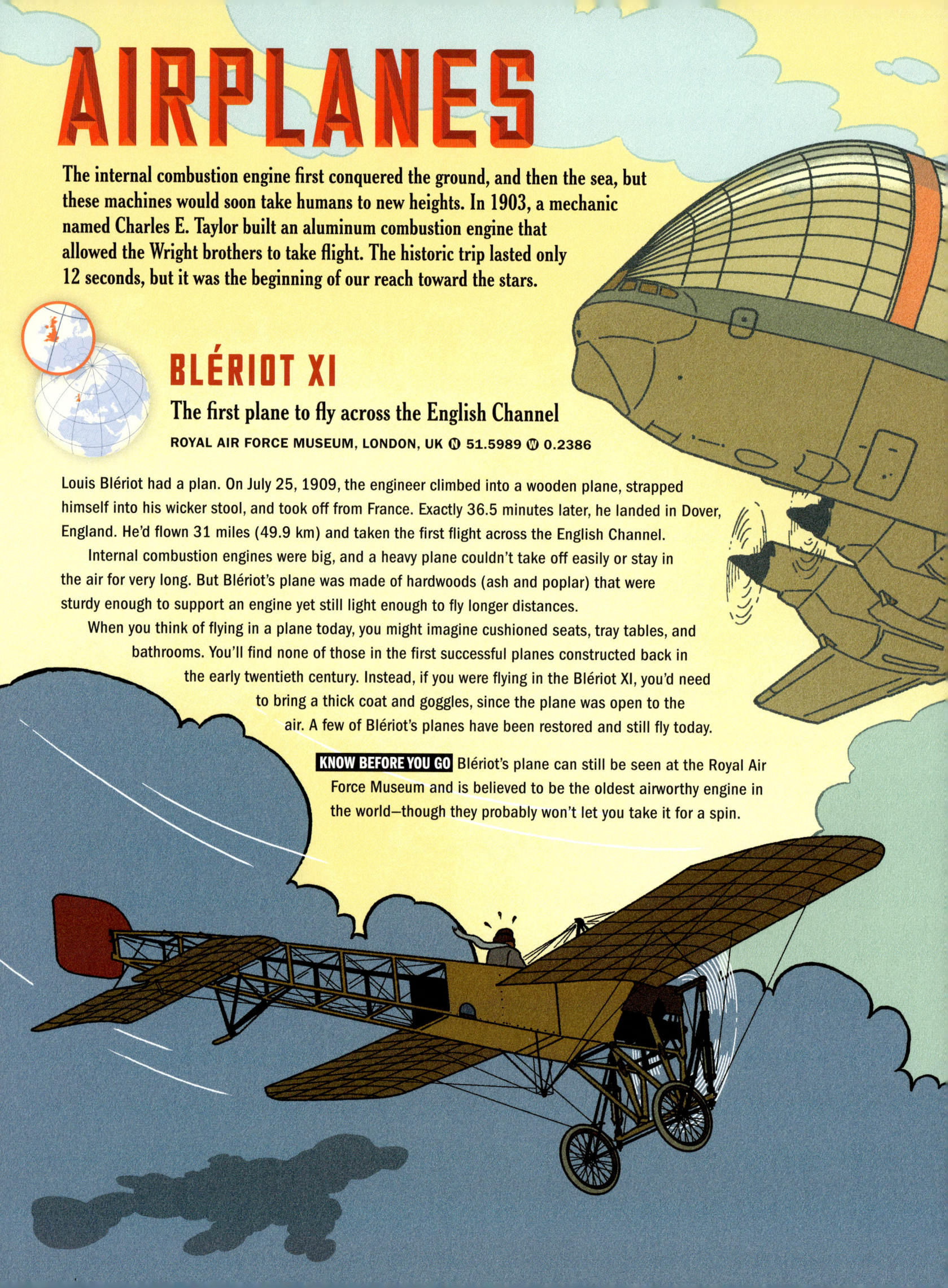

BLÉRIOT XI

The first plane to fly across the English Channel

ROYAL AIR FORCE MUSEUM, LONDON, UK Ⓝ 51.5989 Ⓦ 0.2386

Louis Blériot had a plan. On July 25, 1909, the engineer climbed into a wooden plane, strapped himself into his wicker stool, and took off from France. Exactly 36.5 minutes later, he landed in Dover, England. He'd flown 31 miles (49.9 km) and taken the first flight across the English Channel.

Internal combustion engines were big, and a heavy plane couldn't take off easily or stay in the air for very long. But Blériot's plane was made of hardwoods (ash and poplar) that were sturdy enough to support an engine yet still light enough to fly longer distances.

When you think of flying in a plane today, you might imagine cushioned seats, tray tables, and bathrooms. You'll find none of those in the first successful planes constructed back in the early twentieth century. Instead, if you were flying in the Blériot XI, you'd need to bring a thick coat and goggles, since the plane was open to the air. A few of Blériot's planes have been restored and still fly today.

KNOW BEFORE YOU GO Blériot's plane can still be seen at the Royal Air Force Museum and is believed to be the oldest airworthy engine in the world—though they probably won't let you take it for a spin.

BLÉRIOT XI

SUPER GUPPY

SUPER GUPPY

It's a bird! It's a fish! It's the Super Guppy!

PIMA AIR & SPACE MUSEUM, TUCSON, ARIZONA, USA
Ⓝ **32.1389** Ⓦ **110.8687**

When walking onto the tarmac of the Pima Air & Space Museum, you probably wouldn't expect to see a plane that looks like a fish. But when you need to transport something *really* big, you send for the Super Guppy!

Named after the colorful tropical fish, this bulbous plane can carry cargo too massive to fit into anything else. The cargo bay looks like a cross between a plane and a balloon, and the plane's "nose" opens to an interior 25 feet (7.6 m) wide and 111 feet (33.8 m) long. The Super Guppy is now owned by NASA and has transported the *Orion* space capsule, the spacecraft taking humans back to the moon during the Artemis missions.

How does this plane fly? Four massive turboprop engines, each one with 4,600 shaft horsepower, propel the plane at speeds of up to 300 mph (480 kph). Plenty impressive since it carries up to 45,000 pounds (20,400 kg)—about three *T. rexes'* worth—and there's room for ten *T. rex*-sized containers inside. Dino delivery coming right up!

KNOW BEFORE YOU GO The Pima Air & Space Museum offers an outdoor tour of the Super Guppy. If you're lucky, you might even see this plane with its "mouth" open.

ROCKETS

With the invention of planes, the sky became another world to explore. But to reach past Earth's atmosphere, you need a rocket! A rocket burns materials at much higher temperatures than other combustion engines. That heat gives the rocket the boost required to get into space. In 1902, a Peruvian inventor named Pedro Paulet proposed an idea for a liquid-fuel rocket plane. By 1926, the first modern rocket was launched, and the sky was no longer the limit.

SPACEPORT AMERICA

A launchpad for the people

LAS CRUCES, NEW MEXICO, USA Ⓝ 32.9903 Ⓦ 106.9697

Way out in the New Mexico desert, there is a futuristic building with an eco-friendly curved roof that looks a bit like an alien spaceship. This is where dreams come true—at least for dreamers who have always wanted to go to outer space! At Spaceport America, people are propelled into the stratosphere and beyond as commercial astronauts.

As the world's first commercial spaceport, it is home to several companies that conduct research and launch satellites and people from right here at the complex. It has a runway, a vertical launchpad, and an advanced technology area for trying out new ways of blasting cargo into space. One company called SpinLaunch is trying to whip satellites into the atmosphere using centrifugal force, basically by spinning them around *really* fast and then quickly letting go.

Spaceport America also hosts rockets from NASA and the US military and space planes from many private companies. A superlarge assembly building allows engineers to build rockets and satellites on-site, so transportation to the launchpad is easy.

KNOW BEFORE YOU GO On your tour of the spaceport, be sure to try the G-Shock Trainer. It's a rapidly spinning machine that re-creates the g-forces astronauts feel when they launch.

THE JET PROPULSION LABORATORY

Before NASA there was JPL.

PASADENA, CALIFORNIA, USA Ⓝ 34.2000 Ⓦ 118.1717

In a quiet corner of Pasadena, California, away from the bustling city of Los Angeles, sits one of the most important hubs of space exploration history. The Jet Propulsion Laboratory, or JPL, is where the story of US spaceflight begins.

Frank Malina grew up reading Jules Verne's science fiction books and dreaming of one day traveling to other planets. He was just 23 when he started a groundbreaking mission. Along with fellow engineers Ed Forman and Jack Parsons, Malina began creating powerful rockets for use in jets in the 1930s. Their experiments were extremely dangerous. The group also had a secret: Their true aim was to take the United States to the stars, at a time when many thought the idea was ridiculous. Malina's team launched their first rocket to reach space in 1945. It worked! Too well, in fact. The military wanted the technology to build bombs, but Malina refused and would soon resign from the JPL. Later, in 1958, the JPL successfully launched the United States' first satellite, Explorer 1, marking the birth of the US space program.

KNOW BEFORE YOU GO JPL continues to push the boundaries of space exploration by sending robotic missions to distant planets. You can visit JPL on its free public tours.

SATELLITES

On October 4, 1957, the Soviet Union launched a rocket carrying Sputnik, the first satellite, into space. Rockets have brought astronauts to the moon and telescopes into orbit to peer into deep space, but we interact with satellites on a regular basis. Today there are more than 11,000 satellites orbiting Earth. Over 3,000 are space junk, more than 6,000 are for communication, 130 or so are for science, and about 150 tell you where you are in the world. Space is a busy place!

THE GLOBAL POSITIONING SYSTEM

The little blue dot that tells you where you are in the world

ORBITING THE EARTH

Far above your head—12,550 miles (20,200 km) above it, to be exact—there are 31 navigation satellites. Each one is between the height of an adult giraffe (17 feet or 5 m) and nearly the length of a semi-truck (60 feet or 18 m) and whizzing through space at about 7,000 mph (11,265 kph). Each satellite carries a clock with nearly perfect accuracy, and together they all tell you where you are.

This is the Global Positioning System (GPS). The satellites follow six equally spaced orbital paths around Earth. Each satellite looks like a blocky refrigerator with wings. The "wings" are large solar panels providing power, while the rectangular-shaped "block" contains the computer equipment and a radio that sends signals back to Earth.

In a way, a GPS satellite is just a fancy clock. Each one has an atomic clock (accurate down to three-billionths of a second) that records the precise instant the satellite transmits a radio signal to Earth. The satellites are constantly beaming these signals to our planet. It's like shouting the mathematical equivalent of "I'm here now! . . . Now here!"

How do these "shouts from space" tell you where you are? Your phone receives a signal from four individual GPS satellites. The receiver in your phone compares the time it takes for a signal to arrive from each of the satellites to calculate an accurate position. Three of the satellites are for directionality (like the x-, y-, and z-axes on a 3D graph) and the fourth is to correct for the time difference of your own clock. If your phone clock is off by even a millionth of a second, your GPS could be wrong by a quarter of a mile!

GPS isn't just useful for finding your way. It's superimportant to the government, the military, and business. The 31 GPS satellites used in the United States are government-owned and operated by a US Space Force unit called Space Delta 8. GPS is just one system of positioning satellites out there. China has its own system called BDS, while Europe's is called Galileo. Today, it's harder and harder to imagine us getting around without these navigational systems.

KNOW BEFORE YOU GO To see a GPS satellite up close, check out the National Air and Space Museum in Washington, DC, which has a model of the Block II type. If you want to know when a GPS satellite might be passing overhead, you can track them on several websites.

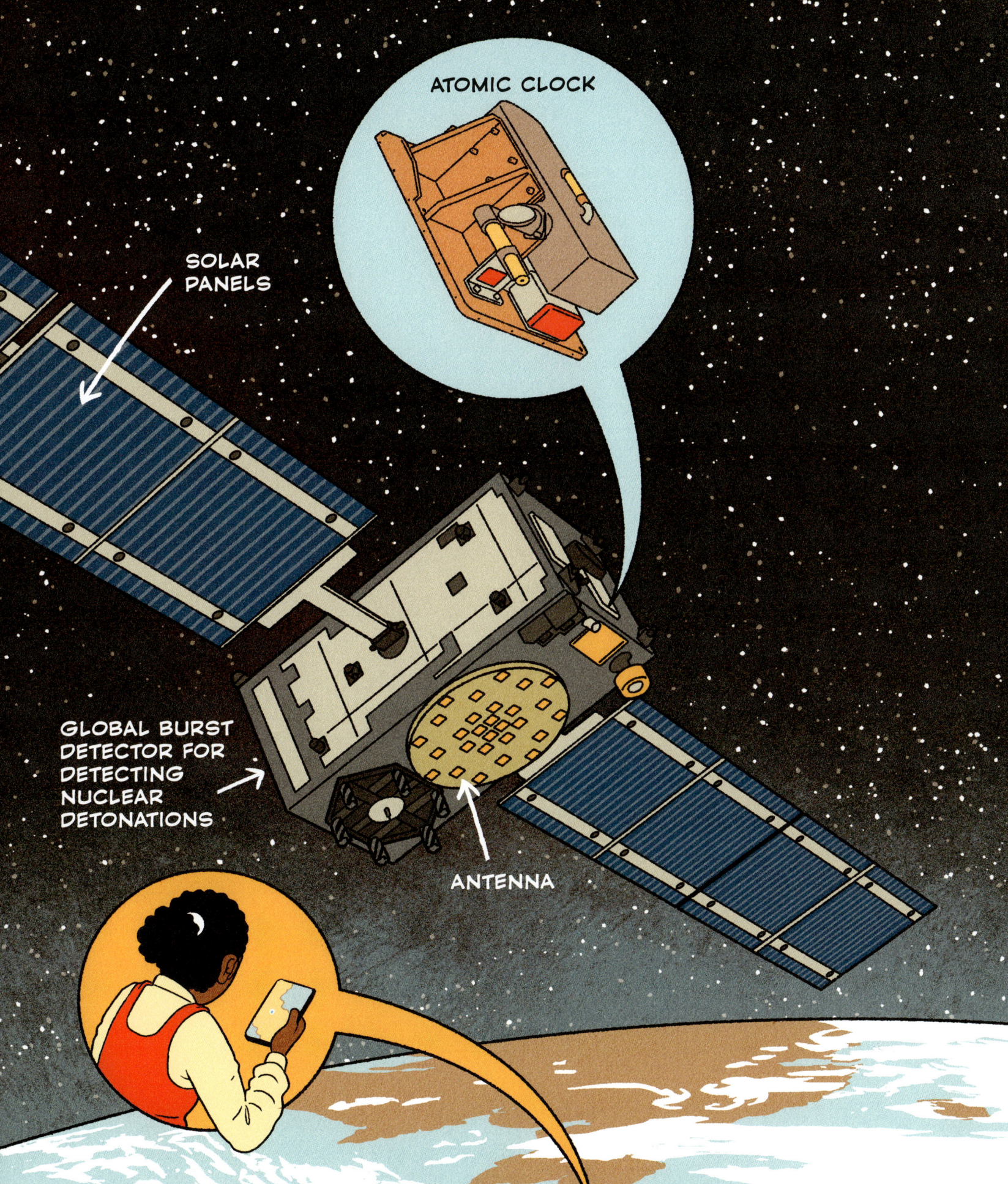

MAPS

GPS satellites can tell you *where* you are, but not *what* is actually there. This has caused some interesting problems! People who followed their phone GPS without looking have ended up driving into a lake or falling down stairs in a park. To know what's really around you, you can rely on a much older technology: maps.

6 LEAGUES IN BETWEEN WHERE THE SUN IS NOT SEEN

6 LEAGUES IN BETWEEN

REGION

GREAT WALL

MOUNTAIN

OCEAN

HABBAN

CITY

OCEAN

BABYLON

URARTU

8 LEAGUES IN BETWEEN

CITY

ASSYRIA

REGION

EUPHRATES

BIT YAKIN

CHANNEL

SWAMP

DER

SUSA

BABYLONIAN MAP OF THE WORLD

The Atlas Obscura of 800 BCE

BRITISH MUSEUM, LONDON, UK Ⓝ 51.5194 Ⓦ 0.1269

Today, maps are on our phones and in our cars. But thousands of years ago, they were on tablets. No, not *that* kind of tablet—clay tablets! The oldest map of the known world was discovered in 1881 by an Iraqi archaeologist near the ancient city of Babylon. At the time, he didn't think much of the small chunk of clay with its curly symbols and lines. But later it was determined to be a nearly 2,800-year-old map of the Babylonian civilization!

This tablet, now in the British Museum, has symbols at the top inscribed in a written language called cuneiform to explain the map. Pictured is the Euphrates River flowing from north to south, with Babylon on the north end of the river. Three different capital cities—Susa, Urartu, and Habban—are also depicted, and the map shows a traveler how to get to these places. The cuneiform also describes mythical beasts and great heroes that lived in those areas. It's a map *and* tour guide all in one!

KNOW BEFORE YOU GO Among the many animals listed on the backside of the map are a lion-bird and a scorpion-man! The map is in room 55 of the British Museum, also home to the Nimrud Lens.

CERRO TOLOLO INTER-AMERICAN OBSERVATORY

The largest map of space ever made

LA SERENA, CHILE Ⓢ **30.1697** Ⓦ **70.8065**

At the top of a 7,200-foot-high (2,194 m) mountain sit four white domes. They seem weirdly out of place amid the beautiful greenery of Chile. Maybe that's because they are there to create a map of the planets and stars beyond our world.

Rumbling noises signal the opening of the giant telescopes sitting inside the Cerro Tololo Inter-American Observatory. The astronomers are using their Dark Energy Spectroscopic Instrument (DESI), a kind of telescope that detects different types of light and energy, to build the largest-ever 3D map of the sky.

To create this map, astronomers used three major telescopes, one of which was the Cerro Tololo, to take pictures of the night sky for 1,405 days. More than 200 observers and researchers around the world contributed to the petabyte (1 million gigabytes) of data collected to make this 10-trillion-pixel image. It would take more than 830,000 camera phones to collect enough data to create an image that big. What can we do with this space map? Figure out amazing places to point telescopes at and investigate next!

KNOW BEFORE YOU GO The Cerro Tololo Inter-American Observatory offers free public tours, and you can also see some of the observatory's amazing images of deep space online.

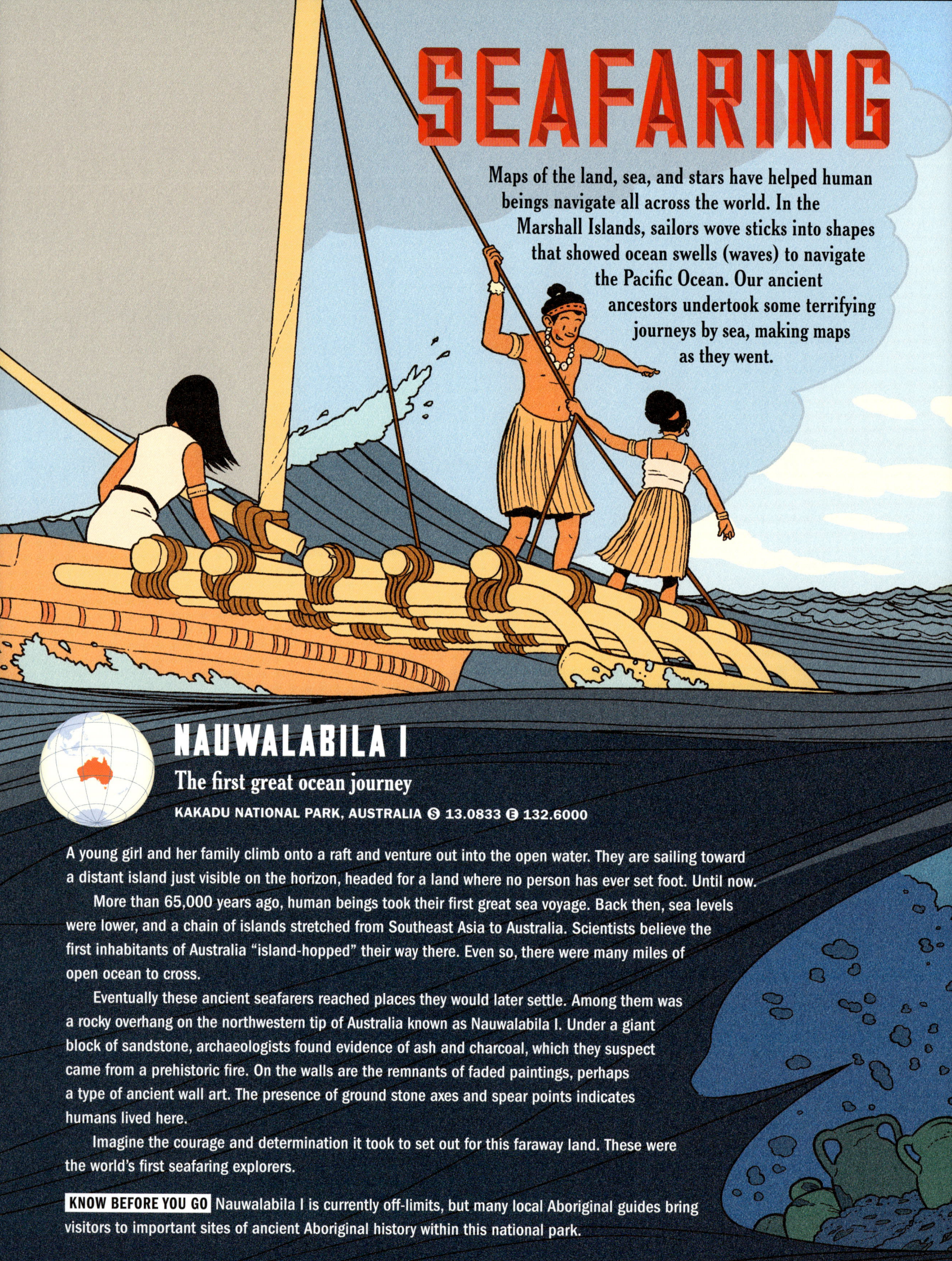

SEAFARING

Maps of the land, sea, and stars have helped human beings navigate all across the world. In the Marshall Islands, sailors wove sticks into shapes that showed ocean swells (waves) to navigate the Pacific Ocean. Our ancient ancestors undertook some terrifying journeys by sea, making maps as they went.

NAUWALABILA I

The first great ocean journey

KAKADU NATIONAL PARK, AUSTRALIA ⓢ 13.0833 Ⓔ 132.6000

A young girl and her family climb onto a raft and venture out into the open water. They are sailing toward a distant island just visible on the horizon, headed for a land where no person has ever set foot. Until now.

More than 65,000 years ago, human beings took their first great sea voyage. Back then, sea levels were lower, and a chain of islands stretched from Southeast Asia to Australia. Scientists believe the first inhabitants of Australia "island-hopped" their way there. Even so, there were many miles of open ocean to cross.

Eventually these ancient seafarers reached places they would later settle. Among them was a rocky overhang on the northwestern tip of Australia known as Nauwalabila I. Under a giant block of sandstone, archaeologists found evidence of ash and charcoal, which they suspect came from a prehistoric fire. On the walls are the remnants of faded paintings, perhaps a type of ancient wall art. The presence of ground stone axes and spear points indicates humans lived here.

Imagine the courage and determination it took to set out for this faraway land. These were the world's first seafaring explorers.

KNOW BEFORE YOU GO Nauwalabila I is currently off-limits, but many local Aboriginal guides bring visitors to important sites of ancient Aboriginal history within this national park.

ANTIKYTHERA SHIPWRECK

An ancient shipwreck still revealing secrets

AEGEAN SEA Ⓝ 35.8814 Ⓔ 23.3176

Sometime around 70 BCE, off the Greek island of Antikythera, things went very, very wrong. A Roman ship packed with expensive goods sank to the bottom of the sea, taking passengers and crew with it. The ship wasn't seen again until 1900, when Greek sponge divers in big brass- and copper-helmeted diving suits accidentally discovered the wreckage.

What survived the shipwreck? The valuable goods inside included bronze and marble sculptures of horses and Roman gods, a mysterious mechanism for predicting astronomy, and *lots and lots* of jars. Yep, the ship carried hundreds of jars called amphorae. These clay vessels held things like wine, olive oil, and wheat. The jars could be stacked neatly on top of one another, which allowed the Romans to pack and ship all kinds of goods across their empire.

So what caused the Antikythera shipwreck? Bad weather? Bad maps? We don't know for sure, but maybe it was carrying just one too many jars!

KNOW BEFORE YOU GO If you don't feel like diving to the ocean bottom, you can see the Antikythera astronomical mechanism and some of the ancient marble horse sculptures at the National Archaeological Museum in Athens.

CERAMICS

As ceramic (hardened clay) vessels like ancient Roman amphorae made their way onto seafaring ships to transport goods, a new sort of global trade was born. By the 1600s, ceramics themselves had become the focus of seafaring trade as Europe was gripped with "porcelain fever." Rich aristocrats clamored for ceramics from China, but only the Chinese knew how to make the ultrasmooth ceramic called porcelain.

JINGDEZHEN IMPERIAL KILN MUSEUM

The hidden city where fine china was made

JINGDEZHEN, CHINA Ⓝ 29.2951 Ⓔ 117.2069

Slurp a long sip of hot chocolate from your favorite mug and . . . *ouch!* Sometimes your drink is a bit too hot to handle—but the heat required to make the mug itself would be more than enough to vaporize you.

To create that ceramic mug, potters mix clay, bits of rock, mineral dust, and water. They then mold the mixture into shape and place it in a special furnace called a kiln. Kilns are typically made of bricks and can be heated up to 2,200°F (1,204°C)!

For more than 1,700 years, the giant kilns in Jingdezhen, China, have produced a type of ceramic called porcelain. What made porcelain so special that the whole world wanted it? Like ceramic, it's made of clay, but it has crystalline rocks mixed in. Heat it in a kiln, and the porcelain shines brightly. The process was kept top secret—so secret that traders were forbidden from even entering the city. The method remained under wraps until 1708, when an imprisoned alchemist finally cracked the recipe.

KNOW BEFORE YOU GO At the museum, you can walk through a maze of massive kilnlike structures. They aren't the original ones, though—after three years, a kiln's bricks break down, so very few ancient kilns remain.

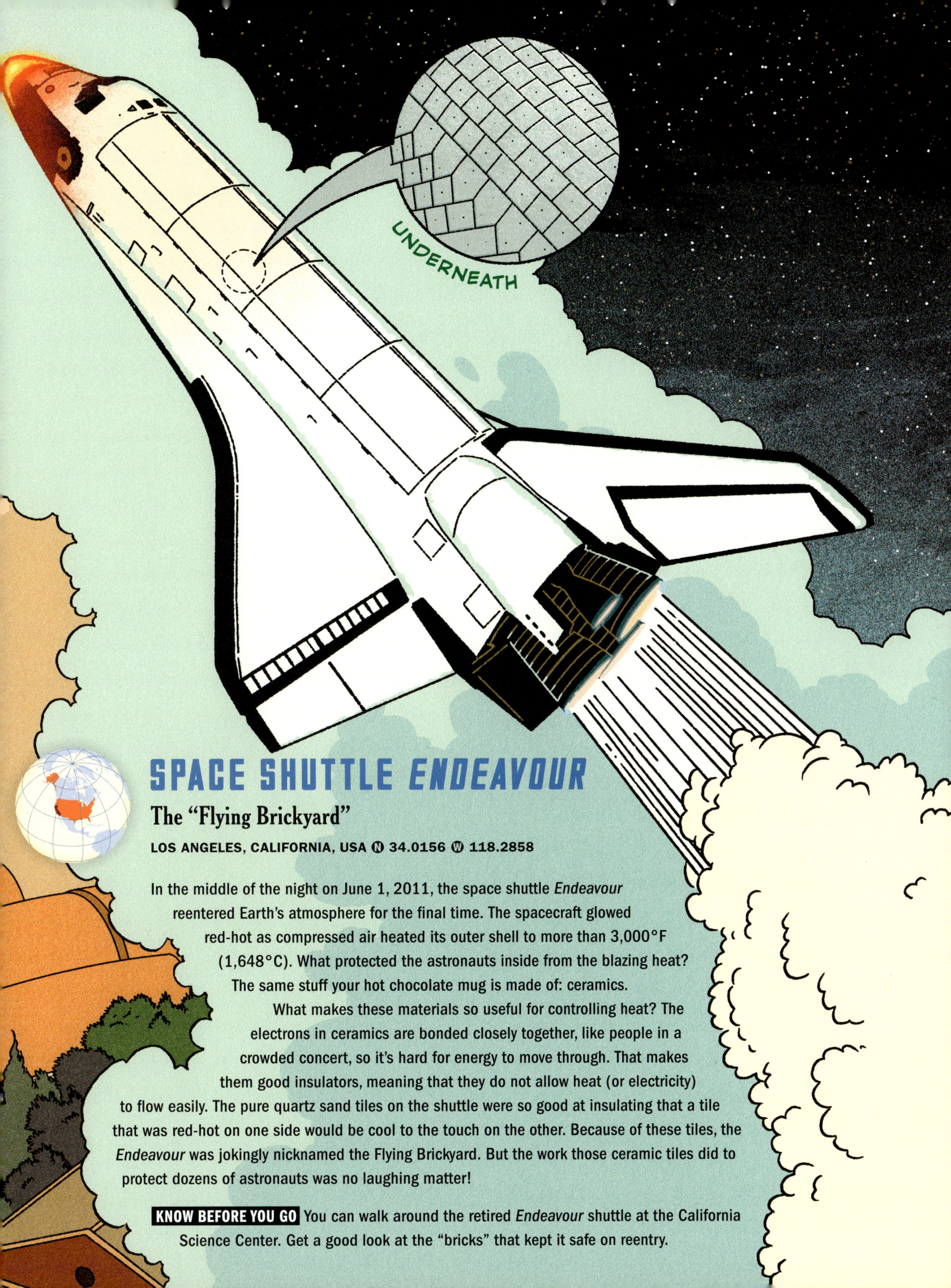

UNDERNEATH

SPACE SHUTTLE *ENDEAVOUR*
The "Flying Brickyard"

LOS ANGELES, CALIFORNIA, USA Ⓝ 34.0156 Ⓦ 118.2858

In the middle of the night on June 1, 2011, the space shuttle *Endeavour* reentered Earth's atmosphere for the final time. The spacecraft glowed red-hot as compressed air heated its outer shell to more than 3,000°F (1,648°C). What protected the astronauts inside from the blazing heat? The same stuff your hot chocolate mug is made of: ceramics.

What makes these materials so useful for controlling heat? The electrons in ceramics are bonded closely together, like people in a crowded concert, so it's hard for energy to move through. That makes them good insulators, meaning that they do not allow heat (or electricity) to flow easily. The pure quartz sand tiles on the shuttle were so good at insulating that a tile that was red-hot on one side would be cool to the touch on the other. Because of these tiles, the *Endeavour* was jokingly nicknamed the Flying Brickyard. But the work those ceramic tiles did to protect dozens of astronauts was no laughing matter!

KNOW BEFORE YOU GO You can walk around the retired *Endeavour* shuttle at the California Science Center. Get a good look at the "bricks" that kept it safe on reentry.

SMELTING METAL

More than 7,000 years ago, a potter paints her clay jar with a bright green ground-up rock called malachite. When she puts her jar in the kiln, the malachite melts, producing small droplets of copper. This ancient ceramicist has just accidentally discovered smelting—the process of removing metal from rock by melting it out. From farming tools to fighting tools, it was goodbye to the Stone Age and welcome to the metal millennia.

ÖTZI THE ICEMAN'S AXE

The axeman cometh.

SOUTH TYROL MUSEUM OF ARCHAEOLOGY, BOLZANO, ITALY
Ⓝ 46.4999 Ⓔ 11.3495

It was late spring or early summer some 5,300 years ago when the hunter-gatherer we call Ötzi (named after the Ötztal Alps mountains in Italy) made his way across a glacier. He carried food and tools, a bow and arrow, his dagger, and a copper axe. Unfortunately, it wouldn't be enough to survive. Someone snuck up and shot Ötzi in the back with an arrow. Over time, ice swallowed his body and preserved it.

About 5,270 years later, a mountain climber in the Alps discovered Ötzi the Iceman, complete with his clothes and axe intact. The copper-edged axe-head was still attached to its wooden handle by leather straps. Ötzi's axe is one of the oldest examples of metal smelting we have found. It also turned out to have been smelted from copper rocks found far away in southern Italy. Ötzi and his axe give us an incredible look into the world of ancient hunter-gatherers and just how advanced their trade routes and toolmaking were, even five millennia ago.

KNOW BEFORE YOU GO You can still see Ötzi's 5,300-year-old mummy—and his axe—at the museum. Ötzi is stored in a special refrigerated chamber and viewed through a small window.

HAYA METAL SMELTING

Ancient steel cooked in termite mound kilns

KAGERA REGION, TANZANIA Ⓢ 1.6252 Ⓔ 31.6200

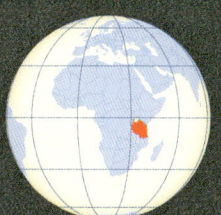

Nearly 2,000 years ago, in what is now Tanzania, the Haya people were making some of the hardest metal on the planet. They used what was at their disposal: towering termite mounds. They did so for good reason; the dirt, dung, and termite saliva that formed the mounds could withstand extreme heat.

Within these mound-shaped kilns, the Haya heat-charred swamp reeds, mixed iron with burnt wood, and blew in air to create a fire hot enough to make carbon steel. This metal was more durable than others available at the time and might have made useful tools and weapons. Today we use carbon steel in blades, skyscrapers, and many other superstrong things.

The Haya were way ahead of their time. It would take another 1,800 years or so before American inventors discovered how to mass-produce metal of this enormous strength.

KNOW BEFORE YOU GO While the ancient kilns have long since crumbled to dust, you can visit the Mushonge Museum at Kamachumu Plateau for a glimpse into what ancient Haya life was like.

STEEL

Metal weapons were first made of copper around 6500 BCE, bronze around 1700 BCE, and iron around 1200 BCE. The harder blade won the fight, and humans kept searching for something even stronger and more durable. They determined that mixing iron and carbon could create an incredibly powerful metal: steel.

BIZEN OSAFUNE SWORD MUSEUM

Just a slice of life for a samurai

SETOUCHI, JAPAN Ⓝ 34.7225 Ⓔ 134.1089

A flash of steel. A movement so quick, you barely see it. The samurai slides his sword back into his belt as his foe falls to the ground.

Samurai were a respected class of honor-bound warriors who arose in Japan in the twelfth century. They were tasked with protecting feudal lands, which they did extremely well with their swords. But the samurai sword is more than just a weapon—it's both an iconic symbol of Japan and also a flawlessly engineered piece of technology.

At the Bizen Osafune Sword Museum, swordsmiths use traditional methods to forge their craft. First, the searing heat of the tatara (clay furnace) envelops an iron-sand-carbon mixture, transforming it into tamahagane, a special kind of steel. After cooling, the furnace is broken and the tamahagane is reserved. A swordsmith then heats, hammers, and folds the steel to remove all the "slag," or impurities. After hundreds of hours of repeatedly folding, heating, and hammering, the sword is shaped into its famous curve and polished to a metallic sheen. It is now ready to be used by a deadly warrior.

KNOW BEFORE YOU GO To see the swordsmiths forge an actual steel sword, visit the museum on the second Sunday of the month.

BURJ KHALIFA

The steel spine of the world's tallest building

DUBAI, UNITED ARAB EMIRATES Ⓝ **25.1972** Ⓔ **55.2742**

Stand at the edge of the Burj Khalifa's 1,821-foot-high (555 m) observation deck and look down. The birds fly *below* your feet. You can see the *top* of a cloud. The thing keeping you from crashing down in a pile of rubble is the incredible power of steel.

The 163-story-tall skyscraper is the tallest on the planet and was completed in 2009. It relies on about 43,000 tons of steel to stand up. The steel is threaded through the reinforced concrete core to form the "skeleton" of the building.

Why use steel? The concrete in the Burj Khalifa has a fatal flaw. Concrete can stand up straight under the pressure of hundreds of floors in a building, but it's not so great when it comes to twisting or bending. That's where the steel rebars (rods) come in. Without steel inside the concrete, the Burj Khalifa would tear itself apart in a windstorm. Not something you want to think about when standing on that observation deck!

KNOW BEFORE YOU GO If you aren't afraid of heights, give the highest observation deck in the world a look. It's on the 148th floor.

IN 2023 TWO FRENCH CLIMBERS CLIMBED THE BURJ. IT TOOK NINE HOURS!

SENSORS ARE PLACED THROUGHOUT THE BUILDING TO MEASURE MOVEMENT.

THE ELEVATORS GO 22 MPH (13.7 KPH) AND ARE SOME OF THE FASTEST IN THE WORLD!

EIFFEL TOWER ENDS ABOUT HERE.

THE SECOND INDUSTRIAL REVOLUTION

The Haya and Japanese forged individual pieces of steel thousands of years ago. The Chinese learned to make larger amounts by blasting air through molten iron. It wasn't until the mid-1800s that inventors in the US stumbled on (or stole) a similar solution. The US soon began creating hundreds of thousands of tons of steel to build bridges, train rails, and telegraph cables. During this Second Industrial Revolution (1840–1914), the world was rebuilt in metal.

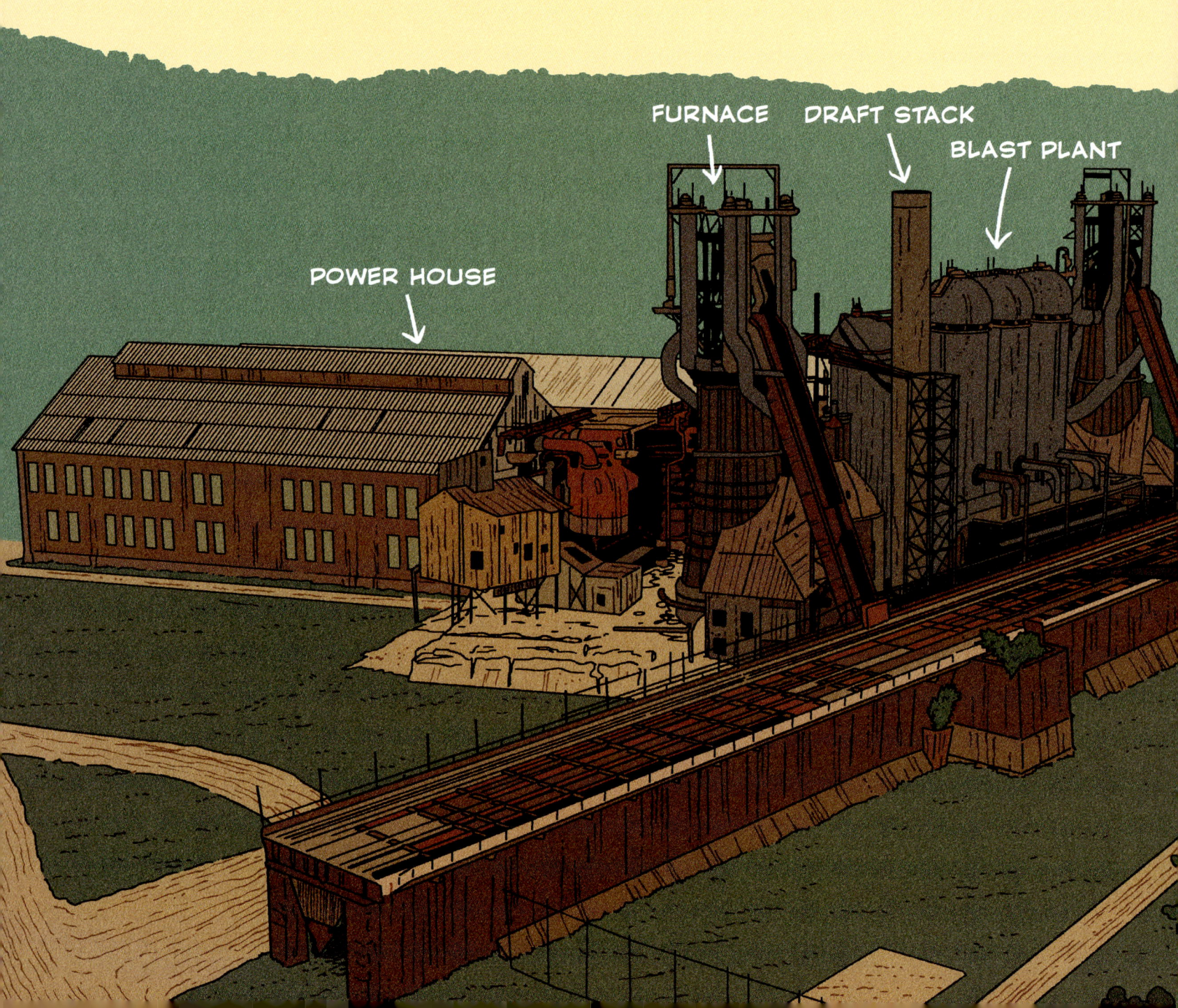

FURNACE

DRAFT STACK

BLAST PLANT

POWER HOUSE

CARRIE BLAST FURNACES

Where rivers of melted steel once flowed

PITTSBURGH, PENNSYLVANIA, USA Ⓝ 40.4131 Ⓦ 79.8901

Rivers of glowing molten metal pouring out of buckets the size of a building. Sparks flying in all directions. Furnaces heated to 3,000°F (1,650°C). This was what working in a late nineteenth-century blast furnace was like.

On the banks of the Monongahela River stand two massive, rusted towers rising 92 feet (28 m) high over Pittsburgh, Pennsylvania. These towers are all that remain of the Carrie Blast Furnaces. From 1884 to 1982, these blast furnaces pumped out more than 1,200 tons of iron ore (the metal used in steel) per day.

The First Industrial Revolution marked the shift from agriculture to factories and steam power. The Second Industrial Revolution ushered in an age of trains, bridges, telegraphs, planes, and skyscrapers. The speed of transportation allowed people to travel farther than ever before and shifted life into cities. Huge amounts of steel were needed to build the new cities, rail lines, and businesses that sprang up. This was the Carrie furnaces' job.

It's not a coincidence that these furnaces are located near a river. For every ton of iron created, 4 tons of coke (preheated coal), iron ore, and limestone were discarded. These by-products were so hot that more than 5 million gallons (18.9 million L) of water were needed each day to cool and wash them away.

Only furnaces number six and seven remain, but you can tour what's left of the site and even go inside one of the enormous furnaces. Don't forget to walk by the Iron Garden, the site of furnaces one and two, which, not surprisingly, once contained highly acidic soil contaminated by years of iron production. Today, however, beautiful plants and small bushes have reclaimed the land.

KNOW BEFORE YOU GO The site of the Carrie furnaces is now a national landmark and center for artists and ecology. The *Carrie Deer* is a 45-foot-tall (13.7 m) deer sculpture made of recycled parts found on the premises. Or take an urban art tour highlighting 30 years of artistic graffiti.

ORE BRIDGE

WIRES

The Second Industrial Revolution created mass-produced wires, which we used for telegraph lines and electricity. Today, wires are everywhere, yet we hardly notice them. Even with cell phones and Wi-Fi, we still rely heavily on wires to connect our world. Undersea internet cables carry 95 percent of our international calls and info!

TELEGRAPH FIELD

The Victorian internet

VALENTIA ISLAND, IRELAND Ⓝ 51.8931 Ⓦ 10.3899

Before the 1800s, the main way to communicate with someone far away was by letter. But mail traveled slowly, and by the time a letter arrived, it was often too late. The person you were writing to was dead, or the war had already begun!

Telegraphs were a big step forward. They sent an electrical signal through a wire from one station to another, allowing people to communicate much faster. But what if you wanted to talk to someone on another continent? You first had to run a wire across the bottom of the ocean! (The same is true with the internet today.)

In August 1858, the first successful telegraph wire was laid on the ocean floor, running across the Atlantic between Ireland and Canada. The Queen of England, who at the time ruled both Ireland and Canada, used the wire to send a message to the President of the United States. It worked! But not for long. After three weeks, the wire was damaged by salt water. So, the work started again. It took eight years, but finally, a wire that could hold up underwater was laid. Today, hundreds of undersea cables many thousands of miles long tie our world together.

KNOW BEFORE YOU GO A plaque sits at the original site of the telegraph station. As you enjoy the beautiful view, use a phone to call an overseas friend! That call will still be routed through undersea cables, just like it would have been in 1858.

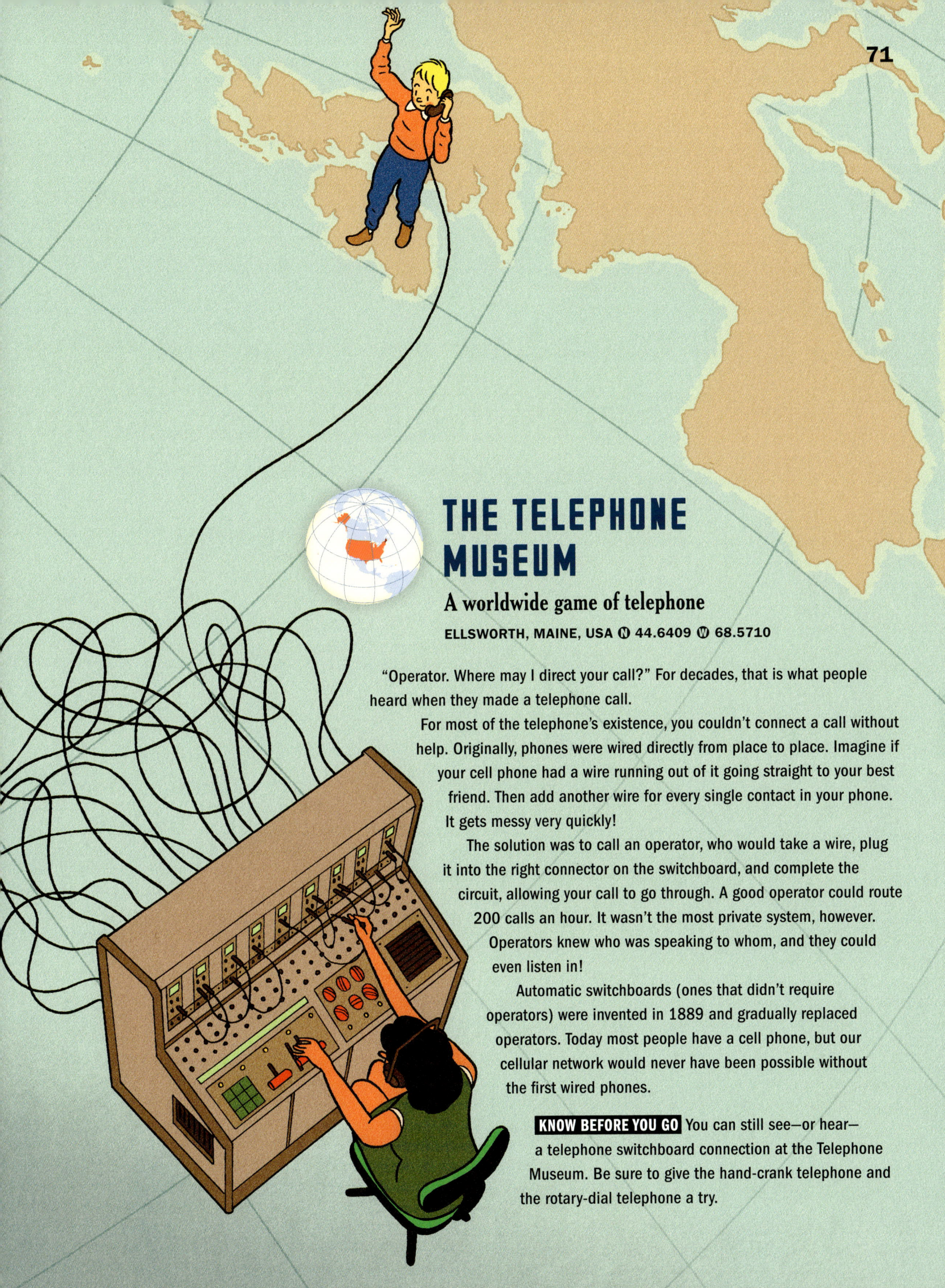

THE TELEPHONE MUSEUM

A worldwide game of telephone

ELLSWORTH, MAINE, USA Ⓝ 44.6409 Ⓦ 68.5710

"Operator. Where may I direct your call?" For decades, that is what people heard when they made a telephone call.

For most of the telephone's existence, you couldn't connect a call without help. Originally, phones were wired directly from place to place. Imagine if your cell phone had a wire running out of it going straight to your best friend. Then add another wire for every single contact in your phone. It gets messy very quickly!

The solution was to call an operator, who would take a wire, plug it into the right connector on the switchboard, and complete the circuit, allowing your call to go through. A good operator could route 200 calls an hour. It wasn't the most private system, however. Operators knew who was speaking to whom, and they could even listen in!

Automatic switchboards (ones that didn't require operators) were invented in 1889 and gradually replaced operators. Today most people have a cell phone, but our cellular network would never have been possible without the first wired phones.

KNOW BEFORE YOU GO You can still see—or hear— a telephone switchboard connection at the Telephone Museum. Be sure to give the hand-crank telephone and the rotary-dial telephone a try.

ELECTRICITY

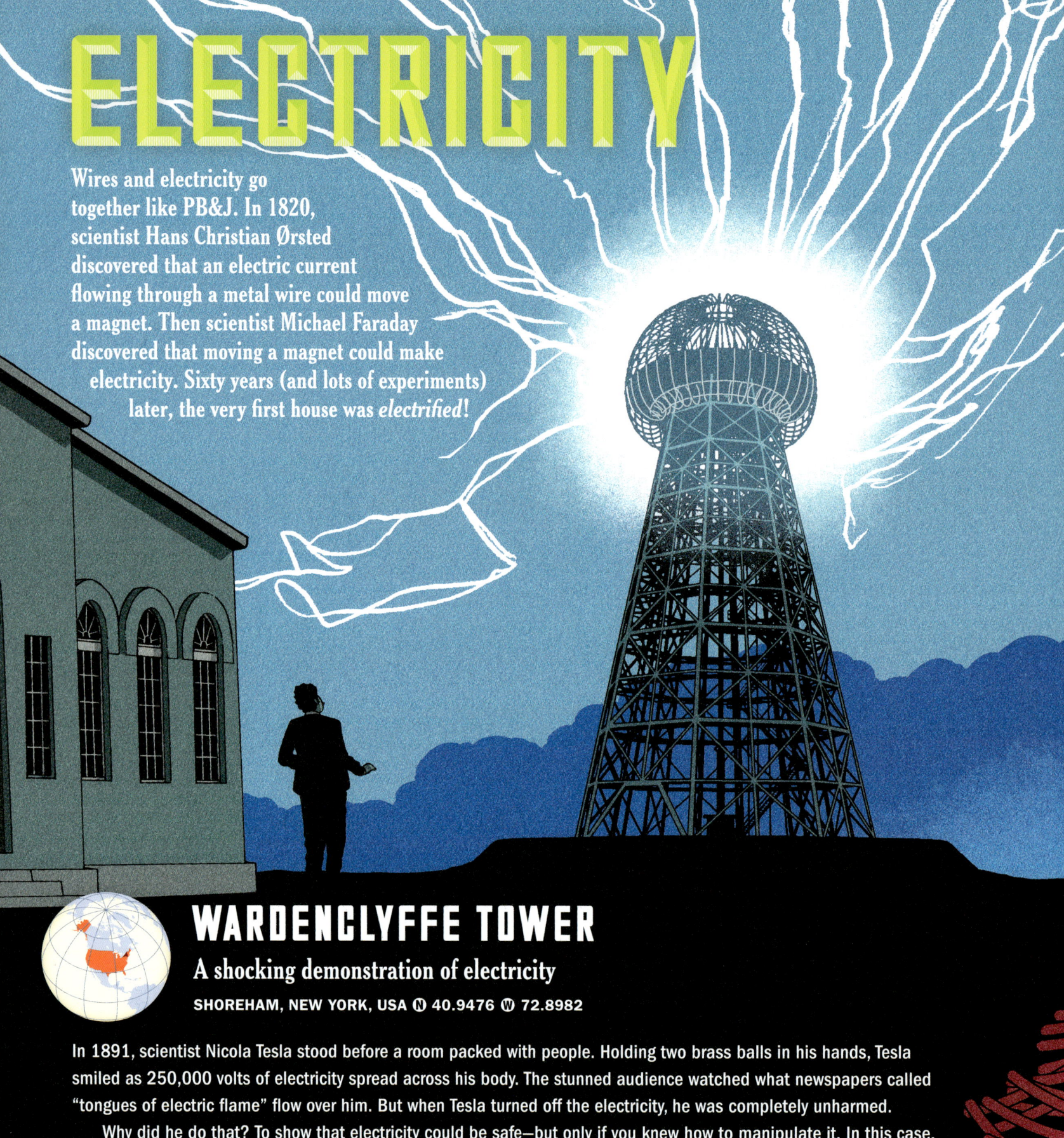

Wires and electricity go together like PB&J. In 1820, scientist Hans Christian Ørsted discovered that an electric current flowing through a metal wire could move a magnet. Then scientist Michael Faraday discovered that moving a magnet could make electricity. Sixty years (and lots of experiments) later, the very first house was *electrified*!

WARDENCLYFFE TOWER

A shocking demonstration of electricity

SHOREHAM, NEW YORK, USA Ⓝ 40.9476 Ⓦ 72.8982

In 1891, scientist Nicola Tesla stood before a room packed with people. Holding two brass balls in his hands, Tesla smiled as 250,000 volts of electricity spread across his body. The stunned audience watched what newspapers called "tongues of electric flame" flow over him. But when Tesla turned off the electricity, he was completely unharmed.

Why did he do that? To show that electricity could be safe—but only if you knew how to manipulate it. In this case, Tesla set the frequency of the electricity so it would pass *over* him, rather than through him. Tesla built the 187-foot-tall (57 m) Wardenclyffe Tower as an extension of this idea. Tesla believed that he could get electricity to travel safely anywhere in the world by using Earth's atmosphere. Wardenclyffe was a prototype meant to turn our planet into a giant electrical circuit to transmit information—like today's phone calls, email, and texts—all with no wires needed.

Tesla never succeeded with his Wardenclyffe wireless electricity project, but his other work transformed our world. Tesla used rotating magnetic fields to create alternating current, the electric power we use in our homes today.

KNOW BEFORE YOU GO Visit the Tesla Science Center, located next to Wardenclyffe, to learn more about one of the most important electrical inventors in the world.

Z PULSED POWER FACILITY

Extreme power with the flip of a switch

SANDIA NATIONAL LABORATORIES, ALBUQUERQUE, NEW MEXICO, USA

Ⓝ 35.0355 Ⓦ 106.5425

In a regular-looking building nestled beneath the hills of Albuquerque, New Mexico, sits the world's most powerful electrical device. With the flip of a switch, it creates a single, short electrical pulse equal to more than all the energy produced by every power plant on the entire planet. Whoa!

The Z Pulsed Power Facility (nicknamed the Z machine) is a huge battery whose job is to create a pulse of power and then "pinch" it down into a container about the size of an egg. As energy is compressed, it becomes much, much stronger. The Z machine's electrical equipment is so powerful it's kept submerged in a gigantic pool of oil to help contain the electricity. Even so, when they fire the machine, bolts of lightning shoot out. Scientists use the Z machine to learn about renewable power, like fusion energy.

For a few nanoseconds, the Z machine produces power a thousand times that of a lightning bolt. But for each pulse, the machine consumes only the energy needed to run a 100-watt light bulb for two days. From small beginnings comes great power!

KNOW BEFORE YOU GO The Z machine is off-limits unless you are a researcher, but Sandia Labs has a great online tour of it. You also can visit other lab facilities like the Gamma Irradiation Facility and Sandia Pulsed Reactor Facility.

ELECTROMAGNETIC SPECTRUM

If you were a bee, you'd detect a type of light that humans can't. Called infrared, this light is a part of the waves of energy known as the electromagnetic spectrum. In 1865, scientist James Clerk Maxwell discovered that electricity, magnetism, and light were all energy waves on this bigger spectrum. From radio waves to X-rays, scientists soon discovered many new wavelengths and many different uses for them.

RÖNTGEN LAB

Where we took the first skeleton selfie

UNIVERSITY OF APPLIED SCIENCES, WÜRZBURG, GERMANY ⓝ 49.7999 ⓔ 9.9319

Imagine if you could suddenly see the bones in your body. It would be shocking! That's exactly what happened to Anna Bertha Ludwig.

In the winter of 1895, a scientist named Wilhelm Röntgen was in his lab in Germany, experimenting with an odd sort of light. Röntgen asked his wife, Anna Bertha Ludwig, to lend him a hand—literally! As Anna held her hand in front of a photographic plate, Röntgen aimed the mystery ray at it, and voilà, an image of her bones appeared on the photographic plate. Anna was so shocked to see her bones she reportedly said, "I have seen my death!" Röntgen had accidentally discovered a high-energy electromagnetic wave that could photograph bones through flesh. He called the mysterious energy an X-ray.

Within a year, doctors were using X-rays to help diagnose their patients. What took longer to realize is that too much X-ray exposure can cause serious radiation damage. Today, that's why you wear a lead vest when you get an X-ray.

KNOW BEFORE YOU GO You can still visit the lab where Röntgen discovered X-rays in Würzburg, Germany. The radiation is long gone, but place your hand near the plate and imagine seeing your bones for the first time like Anna!

EUROPEAN X-RAY FREE ELECTRON LASER

Accelerating the future

SCHENEFELD, GERMANY Ⓝ 53.5890 Ⓔ 9.8290

In an underground tunnel that stretches for 2.1 miles (3.4 km) sits a 17.5-gigaelectronvolt superconducting linear accelerator—say *that* ten times fast! The experimental machine brings electrons to nearly the speed of light. As these electrons zip through a magnetic obstacle course in the tunnel, their up-and-down movements cause X-ray flashes a billion times brighter than the X-rays at your doctor's office. The purpose? These bright flashes allow scientists to see inside things at the atomic level.

Opened in 2017, the European X-Ray Free Electron Laser (XFEL) was the first of its kind, and it remains one of only five in the world. Although it's called a laser, the XFEL is more like an X-ray microscope. By generating more than 27,000 pulses per second, the laser can make a 3D movie of things at the nanoscale, or one-billionth of a meter. Researchers have used this to map the atoms in viruses, which could help find cures for them.

KNOW BEFORE YOU GO The European XFEL offers guided tours for students and has a visitor center called Lighthouse. Then grab a bite to eat at BeamStop in the campus cafeteria.

BROADCASTING

At one end of the electromagnetic spectrum are high-energy X-rays. At the other end are long, slow radio waves. Radio waves have lower energy but travel farther than other wavelengths. This makes them a safer and better wave for long-distance communication. The discovery of radio waves allowed for broadcasting: the sharing of sound and images with millions of people at once.

MARCONI WIRELESS STATION

A lifesaving signal

SIGNAL HILL, ST. JOHN'S, CANADA Ⓝ 47.5705 Ⓦ 52.6818

On a cold December day in 1901, a 27-year-old Italian engineer named Guglielmo Marconi sits hunched inside an abandoned hospital. From his perch on a remote hill in Newfoundland, Canada, Marconi is listening intently to a set of headphones crackling with the noise of the Earth's atmosphere. The headphones are connected to a kite flying 500 feet (152 m) in the air. The kite is acting as an antenna, listening for a signal broadcast from more than 2,200 miles (3,540 km) away in Cornwall, England. There, another engineer is sending radio signals by tapping a telegraph. Marconi hands the headphones to his assistant and asks, "Can you hear anything, Mr. Kemp?"

Click. Click. Click. Those three clicks were Morse code for the letter "S"—the first wireless communication across the Atlantic Ocean. Since radio waves were long and low in energy, they could carry a signal over a great distance, bouncing off the atmosphere and traveling across entire oceans. Only a decade later, the radio operator on the *Titanic* would use Marconi's new invention to call for help. Without that wireless call, it's unlikely a rescue ship would have reached *Titanic* at all.

KNOW BEFORE YOU GO Signal Hill, where Marconi received this transatlantic message, is open to visitors and has a tower, a great view of the ocean, walking trails, and a plaque dedicated to Marconi.

PANTELEGRAPH

The first machine to send an "emoji"

NATIONAL
MUSEUM OF SCIENCE
AND TECHNOLOGY LEONARDO DA VINCI,
MILAN, ITALY Ⓝ 45.4630 Ⓔ 9.1712

Among the many amazing inventions in the Leonardo da Vinci museum in Milan is a small, curved metal object shaped sort of like the Eiffel Tower. It doesn't look like much, but this impressive device once sent pictures electronically from one place to another.

Created in the 1850s by Giovanni Caselli, a monk who loved science, the "pantelegraph" could copy drawings. To do so, it used a pendulum (a swinging weight), an electrical signal, a receiver, and two pieces of paper. The pendulum was fitted with special electronics and hung above a sheet of paper with an image on it. As it swung back and forth, the pendulum could "see" the marks on the paper and create an electric signal. On the other end, a receiver "read" these signals and marked chemically treated paper, producing the very same image. (A 🙂 perhaps?) The pantelegraph marked the beginning of sending images electronically—it was the great-grandfather of both the fax machine (ask your grandparents about those!) and broadcast television.

KNOW BEFORE YOU GO In addition to the pantelegraph, this amazing museum houses Leonardo da Vinci's inventions, a full-scale naval ship, and a spaceship!

LASERS

Charles H. Townes, a scientist who worked on radio signals, wanted to find out what happened if instead of letting energy scatter like in broadcasting, you focused it into a narrow beam. His work led to the birth of the laser, which stands for "light amplification by stimulated emission of radiation."

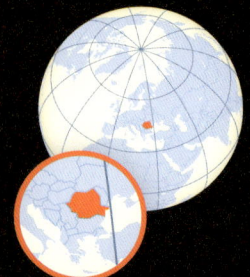

EXTREME LIGHT INFRASTRUCTURE-NUCLEAR PHYSICS

The most powerful garbage disposal on the planet?

MĂGURELE, ROMANIA Ⓝ 44.3500 Ⓔ 26.0498

Imagine operating a laser with the power of a star. Press a button and . . . zzzt! A glowing red light just vaporized part of the universe. Okay, it was just a tiny bit of matter contained within a vacuum-sealed tube. But this laser can really vaporize stuff!

The physicists at the Extreme Light Infrastructure-Nuclear Physics (ELI-NP) lab in Romania operate one of the most powerful lasers on the planet. This 10-petawatt (10 quadrillion watt) laser creates a two-foot-wide (60 cm) red beam focused to roughly one-tenth the power of the sunlight that reaches the Earth. The laser pulse lasts for only around 24 femtoseconds, or 24 quadrillionths of a second. While you may be imagining a death ray, scientists can't actually touch the ELI laser. It's enormous, immovable, kept in strict, sterile conditions, and controlled by computers.

This laser is so strong that it can mimic what happens when a star explodes in a supernova. Why make such a powerful laser? One idea is to use it to vaporize nuclear waste that no one knows what to do with. *Zap!*

KNOW BEFORE YOU GO A couple of times a year, ELI-NP hosts open seminars for students to come and hear about the research being done here. Be sure to sign up in advance!

OPTICAL LATTICE CLOCKS

A tiny laser prison for quickly vibrating atoms

NIST, BOULDER, COLORADO, USA Ⓝ 39.9957 Ⓦ 105.2620

Recipe for creating the most precise clock in the world: (1) Build a freezing-cold prison made of lasers. (2) Capture atoms from the rare metal ytterbium in your prison. (3) Tickle the atoms with energy. (4) Count the wiggles.

Lasers can cut things, measure things, and transmit information. At high-tech US government labs in Boulder, Colorado, researchers use lasers like gears in a clock. The "ticks" in these clocks come from the movement of electrons around an atom's nucleus. Known as optical lattice clocks, they count time in attoseconds, or one billionth of one billionth of one second. Oh, hey—1,000,000,000,000,000,000 (a quintillion) attoseconds just passed.

Why build a laser prison clock? At the National Institute of Standards and Technology (NIST), scientists have the job of defining exactly how long a second is, down to the smallest decimal point. These clocks are so precise, they can measure the subtle variations in gravity that affect time. The laser clocks could run 20 times longer than the universe has existed without ever losing a second. Talk about being on time!

KNOW BEFORE YOU GO Tours of NIST are set up on a case-by-case basis. If you do get close to these clocks, be sure to have your goggles on. A laser in the eye is best avoided!

TIMEKEEPERS

Planets and stars can make useful markers of time, like Earth orbiting the sun (a year) or completing a full rotation (a day). But it was up to humans to define the hour, minute, and second. We can now measure time in trillionths of billionths of a second—so how did we get there? It all started with that big timekeeper in the sky.

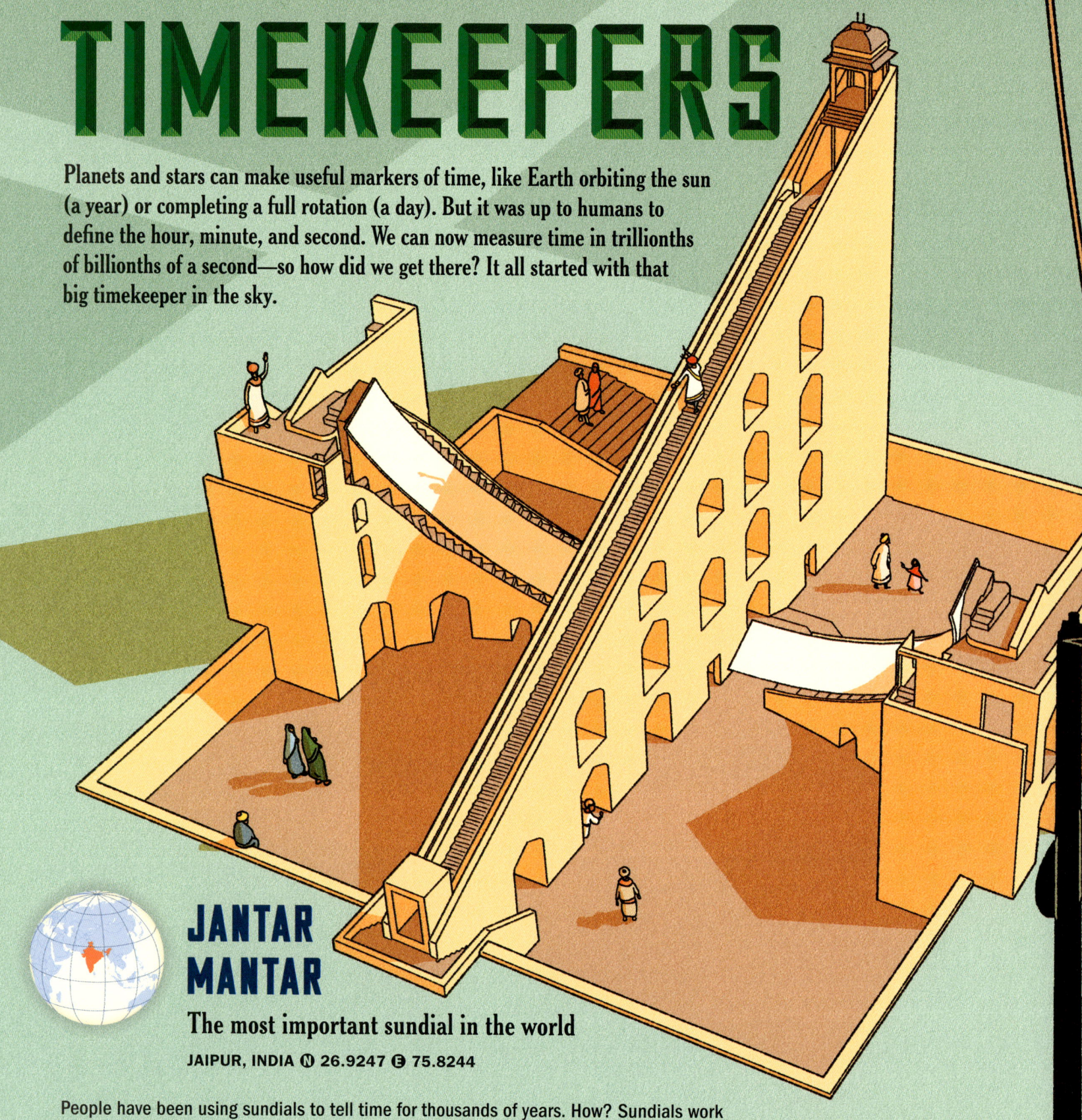

JANTAR MANTAR

The most important sundial in the world

JAIPUR, INDIA Ⓝ 26.9247 Ⓔ 75.8244

People have been using sundials to tell time for thousands of years. How? Sundials work by using a stick to cast a shadow onto a marked circle. The markings indicate hours or minutes. As the day passes (and Earth moves), the shadow from the stick moves around the dial like the hand on a clock, giving you a surprisingly accurate timekeeper.

Now build a towering stone ramp, add a huge curving slab that looks like part of a wicked skate park, and you get a sundial that tells you the time to within two seconds of current timekeeping standards—all with sun and stone. You are at the Vrihat Samrat Yantra (or "great king of instruments") sundial at Jantar Mantar in Jaipur, India. More than 88 feet (26.8 m) tall and wider than a basketball court, it was designed by Maharaja Sawai Jai Singh II, the ruler of the Kingdom of Amber, back in the 1730s.

Jai Singh was extremely interested in astronomy and had more than 20 different instruments built here to measure the "harmony of the heavens." This ancient stone sundial still stands and remains accurate. Just don't try to use it at night!

KNOW BEFORE YOU GO Incredible science tours of the instruments are given at Jantar Mantar. Visit when the sun is out so you can see exactly how the giant sundial works.

SALISBURY CATHEDRAL CLOCK

The oldest clock still going ticktock

SALISBURY, UK Ⓝ 51.0647 Ⓦ 1.7975

A sleepy monk rubs his eyes and heads to the dark tower of Salisbury Cathedral. He winds a piece of heavy rope attached to a weight around several barrels and threads the other end of the rope through a pulley. He pulls the rope until the lead weight is secure against a beam. He repeats the same process with another rope and weight. Then he heads back to bed, his clock-winding job done for the day. As the weights slowly drop, they unwind from the barrels and the bell is struck on the hour every hour.

Built in the 1380s, Salisbury Cathedral's clock is an early example of a mechanical clock. These clocks used a system of interlocking gears and a little mechanism that clicked back and forth to control the descent of the weight and tell the time. These were the first clocks that "ticked." The monks divided daylight into 12 "hour" chunks, so during summer when there was more daylight, the monks just adjusted the clock and made each hour a little longer!

KNOW BEFORE YOU GO The cathedral offers tours to see the clock in action. This six-century-old timekeeper still works and is estimated to have ticked over five billion times.

CLOCKWORK

Keeping exact time requires gears, pulleys, ropes, cogs, weights, springs, and pendulums to all work in precise combinations like, well, clockwork. Ancient clocks were works of both engineering and art. As clockmaking advanced, the instruments got ever more elaborate, complete with life-sized figures, rotating model planets, and other embellishments. These clockwork masterpieces were the first steps toward programmable machines, making clocks the great-great-grandparents of both computers and robots.

HELLBRUNN MECHANICAL THEATER

An entirely automated city inside an unexpected water park

HELLBRUNN PALACE, SALZBURG, AUSTRIA

Ⓝ 47.7614 Ⓔ 13.0606

You are dressed in your finest eighteenth-century clothes: an embroidered long coat, breeches, and a huge lace collar or a heavily detailed dress. You've just finished the most elegant, sumptuous feast. But just before your host invites you to stand up to leave, he pushes a button—*SPLASH*!

A jet of water shoots up from the bottom of your chair, soaking you completely. After a few seconds, the water stops. You stand up, dripping wet, and bow to the Archbishop of Salzburg, who laughs merrily at the joke he has played on his fancy guests.

You are at the Hellbrunn Palace in Salzburg, Austria, where the entire property was designed to perform tricks of engineering and entertainment for the wealthy and royal. Besides the water pranks, the palace was home to one of the most incredible clockwork displays ever created.

Called the Hellbrunn Mechanical Theater, this clockwork was built in 1752. Made up of 163 tiny characters, it illustrated life in the eighteenth century. A tiny guard marches on duty, a group of men construct a building, a farmer pushes a woman in a wheelbarrow, and more. Powered by the movement of water through pipes and a complex system of cogs, gears, and camshafts, each figure moves back and forth or up and down as they interact with one another. It's an entire city of tiny clockwork people.

While Hellbrunn Palace entertained the rich, the Mechanical Theater was built by a local salt miner named Lorenz Rosenegger von Dürrnberg. A mechanical genius, Rosenegger died poor and bitter about this theater. But his technology and creativity were nothing less than incredible.

KNOW BEFORE YOU GO You may want to bring a towel when you visit Rosenegger's theater—you never know when it will be your turn to get splashed.

ROBOTS

A robot is a machine able to perform human movements and abilities. People have been building robots for centuries! During the 1700s and 1800s, inventors created increasingly complex machines, called automatons. Some played piano; others wrote letters or even moved about independently. These uncanny clockwork people were the grandparents of the electronic robots of today.

MAILLARDET'S AUTOMATON

A robotic boy remade from the ashes

FRANKLIN INSTITUTE, PHILADELPHIA, PENNSYLVANIA, USA ⓝ 39.9581 Ⓦ 75.1736

In 1928, a truck pulled up to the Franklin Institute science museum. It was there to drop off what appeared to be an old broken toy the size of a small child. It looked like a burnt metal doll made of gears and complicated brass discs, though the mechanism was missing its legs and seemed mangled beyond use. Curious to understand how it worked, a mechanic named Charles Roberts began tinkering and fixing. When he finally managed to wind it up, the doll sprang to life. These ruined pieces of bronze turned out to be one of the most important early automatons ever created.

Built around 1800 by a Swiss inventor named Henri Maillardet, this mechanical doll was unusual. Its 72 different brass discs, called cams, were carefully sculpted as "mechanical paths." They are the complex memory that, when wound up, allows the clockwork child to draw a garden, cupid, ship, and temple and write three poems before starting over again. When finished with his work, the mechanical boy signs it as "The Automaton."

KNOW BEFORE YOU GO The automaton is on permanent display at the museum. It was also the inspiration for Brian Selznick's kids' book *The Invention of Hugo Cabret* and the movie *Hugo*.

BOSTON DYNAMICS

Robotic creatures that will haunt your nightmares

WALTHAM, MASSACHUSETTS, USA (N) 42.4185 (W) 71.2544

See Spot run. See Spot jump. See Spot use its long, snaking metal neck to pick up a bone. Ummm . . . what?

Spot is not a normal dog. Spot is a doglike robot created in 2016 by the tech company Boston Dynamics. Spot can sit, run, jump, climb over obstacles, flip, and do all sorts of things—except, well, bark. But like a real dog, Spot can be taught to help with search and rescue in emergencies. Spot gathers data by using its sensors (through heat, laser, and sound) and roving areas that are too difficult or dangerous for humans to navigate. The robotics used to make Spot are extremely advanced. They include hydraulics, valves, a custom battery, and tons of those sensors. The "brain" is an advanced computer with artificial intelligence (AI) that allows Spot to learn from its actions and surroundings. Watching Spot move is like looking at a big dog—just creepier and without the slobbery tongue. So far, Spot has surveyed the volcanic ruins of Pompeii, inspected electrical equipment, and even performed a "robo-dog" dance. If the rise of the robots comes, hopefully Spot stays human's best friend.

KNOW BEFORE YOU GO Boston Dynamics offers occasional visits for students and robotics teams. If you have $75,000 to spare, you can have a robot Spot all your own!

PROGRAMMABLE MACHINES

Today's complex electronic robots run on software written by programmers, but the idea of programmable machines was born more than a thousand years ago in the Middle East. Clockwork automatons then furthered the idea. The major breakthrough would later come from an unlikely source. In the early 1800s, a French cloth maker accidentally began a chain reaction of invention that would lead from programmable machines to the advent of modern software.

JACQUARD LOOM

Weaving the future of computing

PAISLEY MUSEUM, PAISLEY, SCOTLAND
Ⓝ 55.8453 Ⓦ 4.4303

Smash! Crash! A small group of people called Luddites stormed through the doors of the loom factory, breaking apart the machines and scaring the workers. Also known as frame breakers, the Luddites were master weavers who designed textile (cloth) patterns. Being a master weaver was a skilled, well-paying job. Until it wasn't.

French weaver Joseph-Marie Jacquard wanted to make fabric faster—and cheaper—so he invented a way to "program" his weaving looms. To write the program, Jacquard punched holes through small metal cards in a specific order. He strung the cards together and fed them through a "card reader" at the top of the loom. These cards were the instructions, telling the mechanized loom how to weave the thread into the programmed pattern. Factory owners used the threat of these programmable machines to force master weavers out of their jobs. The Luddites took notice!

Today, artificial intelligence, the great-great-great-grandchild of these Jacquard looms, can do lots of the work that people currently do. Without making thoughtful choices about technology, who controls it, and who benefits from it, we might find ourselves in the same boat as the Luddites!

KNOW BEFORE YOU GO The recently restored Paisley Museum is home to an antique Jacquard loom. Please don't smash it!

OLD BRASS BRAINS

A system of pulleys smarter than a hundred mathematicians

NOAA FACILITY, SILVER SPRING, MARYLAND, USA Ⓝ 38.9938 Ⓦ 77.0313

In 1910 a US government employee walked up to a 10-foot-long (3 m), 6-foot-high (1.8 m) brass machine and began turning a crank. As the machine's 15,000 components came to life, many whirring, clanking noises filled the room. Known as Old Brass Brains, this machine performed what was once thought impossible: It predicted the movements of the ocean.

Tides, the natural rise and fall of the sea throughout the day, are very complex. They are affected by the gravity of the moon, the wobble of the Earth, the shape of the ocean floor, and much more. Get the tides wrong and your ship might run aground! In the 1860s, mathematicians figured out how to calculate these factors, but it took months to do by hand.

Old Brass Brains (officially "Tide-Predicting Machine No. 2") used a complex system of 37 programmable pulleys to do in a day what would have taken a hundred mathematicians the same amount of time. It was so good that it helped cut the number of shipwrecks in half!

KNOW BEFORE YOU GO Old Brass Brains was in use until 1965, and it's still in working order today. The National Oceanic and Atmospheric Administration (NOAA) occasionally has an in-person open house where you can see it for yourself.

COMPUTERS

Programmable machines were useful, but what if you built a machine that could do *any* kind of calculation? Even before electronics came along, inventors dreamed of computers.

FUGAKU SUPERCOMPUTER

A billion-dollar machine performing red-hot computation

RIKEN CENTER FOR COMPUTATIONAL SCIENCE, KOBE, JAPAN Ⓝ 34.6534 Ⓔ 135.2205

On a small island hanging off the edge of the city of Kobe, Japan, there is a strange-looking building. Sleek and square, its rooftop overflows with white billowy clouds, as if the structure were on fire or housing a nuclear power plant. But those clouds aren't made by the heat of atomic energy. They're the steam created by cooling one of the world's fastest supercomputers. Known as Fugaku, this computer does so many calculations so quickly that the facility gets about as hot as a nuclear reactor. The supercomputer can perform more than 442 quadrillion (that's a thousand trillion) computations per second.

Looking for a way to solve climate change? This supercomputer creates models that show where carbon dioxide levels are the highest in the world and estimates their effects on the planet. This steaming-hot computer could help keep the world cool.

KNOW BEFORE YOU GO The RIKEN center offers tours with an advance application. One day a year on Open Day, anyone can visit some of the advanced labs.

THE DIFFERENCE ENGINE

Ada Lovelace, the first computer programmer

SCIENCE MUSEUM, LONDON, UK Ⓝ 51.4975 Ⓦ 0.1747

Ada Lovelace spent her childhood quietly dreaming about building machines. In 1883, at just 17 years old, she got her chance. That was when she met fellow inventor Charles Babbage and began corresponding with him about mathematics. Babbage had worked on a giant mechanical calculator that he called the Difference Engine. Lovelace was fascinated by Babbage's designs and was among the first to understand their enormous potential.

Lovelace realized that this kind of mechanical computer (using gears instead of circuits) could perform far more than just mathematical calculation. If built big enough, it could solve almost any kind of problem. It could even compose music, Lovelace reasoned, based on its ability to make logical decisions. It just needed the right set of instructions. So, Ada wrote them.

Using her expertise with math and machines, and with Babbage's help, Lovelace wrote what she called a new "language." She created a series of paper punch cards to program the machine. Her work is considered the first computer code, making her the world's very first "coder."

KNOW BEFORE YOU GO You can see a version of the Difference Engine in London's Science Museum. Don't forget to stop and view half of Babbage's brain in a jar. Really!

VIDEO GAMES

Video games began as experiments on early supercomputers. In 1958, a physicist created the first video game for fun, called *Tennis for Two*, featuring a two-dimensional "ball" bouncing between flat paddles. This type of programming, or software, is now used in everything from calculating weather to jumping around in *Fortnite*.

SPACEWAR!

A life-or-death battle made of little white lines

COMPUTER HISTORY MUSEUM, MOUNTAIN VIEW, CALIFORNIA, USA
Ⓝ 37.4142 Ⓦ 122.0774

Five friends sit with hands on their spaceship controls, ready to fight to the death. The five spaceships circle each other, looking for an opening. There can be only one winner. As the ships maneuver around, avoiding a killer star's gravitational pull, they fire off their torpedoes. Boom! Ships explode, and the game starts again.

This is *Spacewar!*, one of the earliest video games, and the first you could play with multiple friends over a network. To play an early version of the game, you had to load the computer program (a long roll of paper punched with holes) into a humming and clacking computer the size of a fridge.

A group of young hackers at Massachusetts Institute of Technology (MIT) created the game in 1962. Its simple graphics of spaceships, a star, and torpedoes became so popular among college students, scientists, and technicians that *Spacewar!* inspired the first arcade games and almost single-handedly kick-started the video game industry.

KNOW BEFORE YOU GO You can still play *Spacewar!* on that giant clacking computer (a PDP-1). The Computer History Museum is full of fascinating exhibits and artifacts, including an unusual half-bicycle, half-computer system called BEHEMOTH.

VERTICAL MOTION SIMULATOR

A moon flight simulator, complete with motion sickness

AMES RESEARCH CENTER, MOUNTAIN VIEW, CALIFORNIA, USA Ⓝ 37.4152 Ⓦ 122.0627

Climb aboard the large, white cube-shaped object attached to the giant moveable platform. Find your cushioned seat and strap in. The two huge video screens in front of you are about to come to life, so grab that joystick, put on your headphones, and get ready for an out-of-this-world piloting experience.

NASA's Vertical Motion Simulator (VMS) is the most realistic space "game" ever created. That's because astronaut pilots use it to practice flying and maneuvering in space. Located in a ten-story-tall building at NASA's Ames Research Center, each of the five customizable VMS machines is connected to rails. These allow the simulators to move up to 60 feet (18.3 m) vertically and 40 feet (12.2 m) horizontally to mimic full motion. Hold on tightly, because the cabs pitch and roll, move forward and backward, and swing from side to side, too. From flying helicopters to operating a Lunar Lander (yes, for landing on the moon), the VMS is as real as simulation gets. Try not to get sick!

KNOW BEFORE YOU GO The research center is closed to visitors, but check out the visitor center nearby in Mountain View to see some space suits, spacecraft, and more. For a kid-friendly flight simulator, try the National Air and Space Museum in Washington, DC.

THE INTERNET

As soon as small networks of computers were set up, people used them to play games together. But every network was separate. A computer scientist named J. C. R. "Lick" Licklider had imagined joining all computers together in what he called the Intergalactic Computer Network. The US government would soon build a small version of the idea called Arpanet, and within a year, people were gaming on it.

THE PACKET RADIO VAN

Driving technology forward—literally

COMPUTER HISTORY MUSEUM, MOUNTAIN VIEW, CALIFORNIA, USA Ⓝ 37.4142 Ⓦ 122.0774

On August 27, 1976, a large computer-filled van with a strangely shaped antenna was parked outside a restaurant. Inside the restaurant, a group of excited shaggy-haired computer engineers ate lunch with long wires running from the van to their table.

Here was the first time *ever* that a mobile network had sent a message to another network. This report was sent via the radio packet network (a very early wireless computer network) to Arpanet, a network located about 500 miles (805 km) away. Thrilled by their success, these engineers from the Stanford Research Institute spent the next several weeks driving around Southern California proving the signals worked no matter where they went. A year later, the "packet radio van" reached satellite networks across Europe. These experiments showed that these packets of information could be shared across a network of networks—what you might call an "inter-net."

KNOW BEFORE YOU GO The Computer History Museum keeps the real van in storage, but it has a model of the van on display alongside the first laptops and personal computers.

THE LONG LINES BUILDING

A secret internet building in the heart of New York City

NEW YORK CITY, NEW YORK, USA
Ⓝ 40.7165 Ⓦ 74.0059

At 33 Thomas Street in Manhattan, New York, is a completely windowless 29-story concrete building originally built for one purpose: to protect America's phone systems from an atomic blast. Designed to house and feed 1,500 people for two weeks in case of a disaster, this strange, looming structure has been called one of the most secure buildings in the US.

Nicknamed the Long Lines Building because of all the long-distance telephone lines connected to it, the building now routes internet traffic as well. While the internet may seem to flow through the air all around us, it actually relies on undersea wires and buildings just like this one.

The Long Lines Building's exact functions have been kept secret. But with so much information passing through, it's been a very attractive target for spies. In 2016, journalists published reports that the National Security Agency (NSA) was using the building in a massive spying operation. Even if the Long Lines Building isn't used for spycraft anymore, this huge windowless manifestation of the internet may still give you the creeps. Famous actor Tom Hanks called it "the scariest building I've ever seen."

KNOW BEFORE YOU GO Even spies can't get inside this ultrasecure building. But if you stroll past just to see it, you might want to walk quickly. Who knows who's watching . . .

BINARY NUMBERS

The internet has its own language. No, not Likes and Shares—it's the language of electricity, in which the only true words are On and Off. This is represented as the "bits" 1 (Electricity On) or 0 (Electricity Off). This language is called binary. 01001000 01101001 is "Hi" in binary. While binary might seem new, we can thank centuries-old mathematicians for inventing the language of machines.

MANGAREVA BINARY SYSTEM

Making your donation count

MANGAREVA, FRENCH POLYNESIA ⓢ 23.1257 ⓦ 134.9797

Sophisticated modern supercomputers use binary as their language. But anthropologists believe that an isolated island in the South Pacific Ocean about 1,000 miles (1,609 km) from Tahiti may have been using this numerical system hundreds of years earlier.

The indigenous people of Mangareva were traders who loved to give large feasts of tribute. The peasants would gather food as a gift to the chief. But each person wanted their share recorded so that the chief would know just how much they gave. How could they keep track?

The Mangarevans developed a unique way of counting. It was a cross between the decimal system we now commonly use (powers of 10, or the numbers 0 through 9) and the binary system (powers of 2, or the numbers 0 and 1). In their system, the first nine numbers were their own numerals, 1 through 9. But the numbers between 10 and 80 were counted in binary form. At 80, they switched back to the decimal system. Little did the Mangarevans imagine that their binary counting system would become the language of machines.

KNOW BEFORE YOU GO Mangareva is a small island, home to around 1,200 people. To get there, first fly to Tahiti; Mangareva is then a four-hour flight from there. It's known for its scuba diving and incredible night skies.

THE STEPPED RECKONER

A calculator invented by the inventor of machine language

GOTTFRIED WILHELM LEIBNIZ LIBRARY, HANOVER, GERMANY Ⓝ 52.3653 Ⓔ 9.7313

Mathematician and philosopher Gottfried Leibniz was a religious man. He looked for God everywhere, especially in numbers. Influenced by the religious ideas of everything and nothingness, as well as Chinese writings on the concept of yin and yang, Leibniz began to develop mathematics that reflected these ideas.

By using only 1 (God, everything, yang) and 0 (the void, nothingness, yin) to represent all numbers, Leibniz saw a path to religious truth. He also saw a mathematical code that might be used for anything: an alphabet of human thought. Leibniz was right. Although he didn't live to see it, binary language became the basis for all digital computers.

Leibniz's love of numbers also led him to create a unique mechanical calculator built with cogs and gears called the stepped reckoner. Invented in 1673, it was a bit more fiddly than today's electronic calculators. But the machine was a marvel of its time, and it could do nearly everything your pocket calculator can.

KNOW BEFORE YOU GO The library has an expansive archive of Leibniz's letters and notes on binary, as well as the only known copy of his stepped reckoner on display.

CODEBREAKING

Gottfried Leibniz was fascinated by the art of creating and solving codes, or cryptography, and his binary system is one way to encode information. You can use similar ideas of replacing one symbol or system with another to translate messages into ciphers (extremely complex codes). But what happens if the encrypted code needs to be deciphered (broken)? That's when you call in the codebreakers.

BOMBE MACHINE

The shack where codebreakers saved the world

BLETCHLEY PARK, UK
Ⓝ 51.9980 Ⓦ 0.7410

In a small wooden building hastily constructed on a large English estate, a young codebreaker named Alan Turing changed the course of World War II.

Sometimes breaking a secret code is literally life and death. During World War II, the Nazis used a machine called Enigma to scramble the letters of the alphabet to encode their messages to each other. The Allies believed that if they cracked Enigma's code, it would help win the war. The problem? The Enigma cipher seemed unbreakable. Enter the brilliant mathematicians secretly working at Bletchley Park.

How did they break the cipher? Using information that Polish codebreakers had gathered, Alan Turing spent months developing a machine called Bombe. It could decode the Enigma cipher by systematically trying many combinations. Some historians believe that Turing's invention shortened the war by as much as two years, saving many lives. Sadly, Turing faced discrimination for being gay and never got the respect or accolades he deserved during his life. He has since become one of the most celebrated computer scientists in the world.

KNOW BEFORE YOU GO Behind the beautiful Bletchley Park mansion are the huts where the Enigma code was broken. Inside, you can still see the Bombe machine run, hear about the people who operated it, and get an explanation of how it worked.

NATIONAL CRYPTOLOGIC MUSEUM

A secret museum of secret codes

ANNAPOLIS JUNCTION, MARYLAND, USA Ⓝ 39.1148 Ⓦ 76.7748

Sitting next door to one of the most secretive buildings in the world, the National Security Agency (NSA) headquarters, is the National Cryptologic Museum. Here you'll find a world of secret codes, impossible ciphers, and everything people have used to covertly communicate with each other. Coding messages can be fun—or deadly serious business.

The oldest encryption machine at the museum was invented by President Thomas Jefferson when he was an ambassador to France. It allowed him to code and decode messages written in French. Called a wheel cipher, its movable wooden disks can be rotated to specific positions to decrypt secret communications. But not all codes require complicated machines. During the 1800s, enslaved people in the American South used patches sewn into quilts to pass along information about the Underground Railroad.

As you walk through the museum halls, you'll be fascinated by the sheer genius involved in creating and deciphering complex codes. Some of the most famous cryptologic advances are on display. Can you crack these codes?

KNOW BEFORE YOU GO You must arrange a visit before going to the museum. Inside, you can learn how everything from Chinese smoke signals to Greek torches were used to transmit secret codes!

LINE UP THE LETTERS SPELLING *CRYPTOLOGY* TO FIND THE SECRET MESSAGE.

QUANTUM COMPUTERS

We all rely on cryptography to keep our information safe. Unfortunately, the codes we use for encryption may one day be worthless. A new type of computer called a quantum computer might crack in minutes codes that would currently take the most powerful supercomputer thousands of years to solve.

CHINA'S QUANTUM NETWORK

Using light and physics to keep your secrets

HEFEI, CHINA Ⓝ 31.8206 Ⓔ 117.2273

Keeping online information private is no joke, and a powerful quantum computer could potentially crack computer security without breaking a sweat. Especially since these machines are cooled to absolute zero (about −459°F or −273°C).

Instead of using binary code like traditional computers, quantum computers use the properties of electrons' spin or the direction of light (photons). Old-school computers rely on bits (1s and 0s), but the bits in a quantum computer are more like waves in an ocean. They can be everything and anything *between* the 1 and the 0. Get the waves to add and cancel each other out in the right ways, and out comes a message in a bottle!

That's probably why China has developed its own supersafe quantum network in the city of Hefei. In these networks, messages are encoded by a quantum system and sent as light pulses through "glass wires" known as fiber-optic cables. If you try to peek at them, the photons change, messing with the message and alerting the sender to snooping. Many consider a quantum network to be "unhackable."

KNOW BEFORE YOU GO Hefei's quantum internet requires 712 miles (1,147 km) of cables. Visit the city and there is a chance you'll be standing right on top of it!

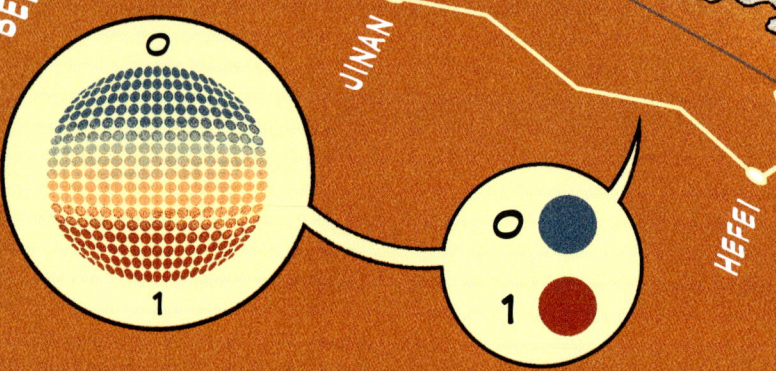

IBM'S QUANTUM LAB

Take a ride through the multiverse.

YORKTOWN HEIGHTS, NEW YORK, USA Ⓝ **1.2102** Ⓦ **73.8030**

You can't be in two places at once, and you can't send a message light-years away instantaneously. Right? Well, quantum mechanics laughs in the face of these limits.

Hanging from the ceiling at IBM's Quantum Lab is a tangle of brass and steel wires. It looks like a cross between an upside-down wedding cake and a chandelier. Mostly it's a refrigerator, meant to cool a handful of atoms to temperatures a thousand times colder than space. This is a quantum computer.

IBM's new Condor quantum computer has 1,121 quantum bits, or qubits, a basic unit of quantum information. Each qubit doubles the computer's power. A quantum computer with 300 functioning qubits would, in theory, be able to perform more calculations than there are atoms in the visible universe! In practice, it's a bit more complicated to make these computers work. But quantum computers have already been used to model magnetic properties and networks similar to the human brain. Some physicists believe that the reason quantum computers work is because the qubits are vibrating between parallel universes. Into the multiverse we go!

KNOW BEFORE YOU GO IBM operates multiple campuses and quantum research facilities in upstate New York, though none offer regular tours.

PARTICLE ACCELERATORS

Quantum computers depend on some strange physics. We never could have built a quantum computer without first understanding the movements and spins of subatomic particles (bits of matter that make up atoms). To help with that, we needed to build giant machines to smash stuff together! By speeding particles to near the speed of light and crashing them together so they explode into their smallest pieces, we can peek inside them. We're still learning with every smash.

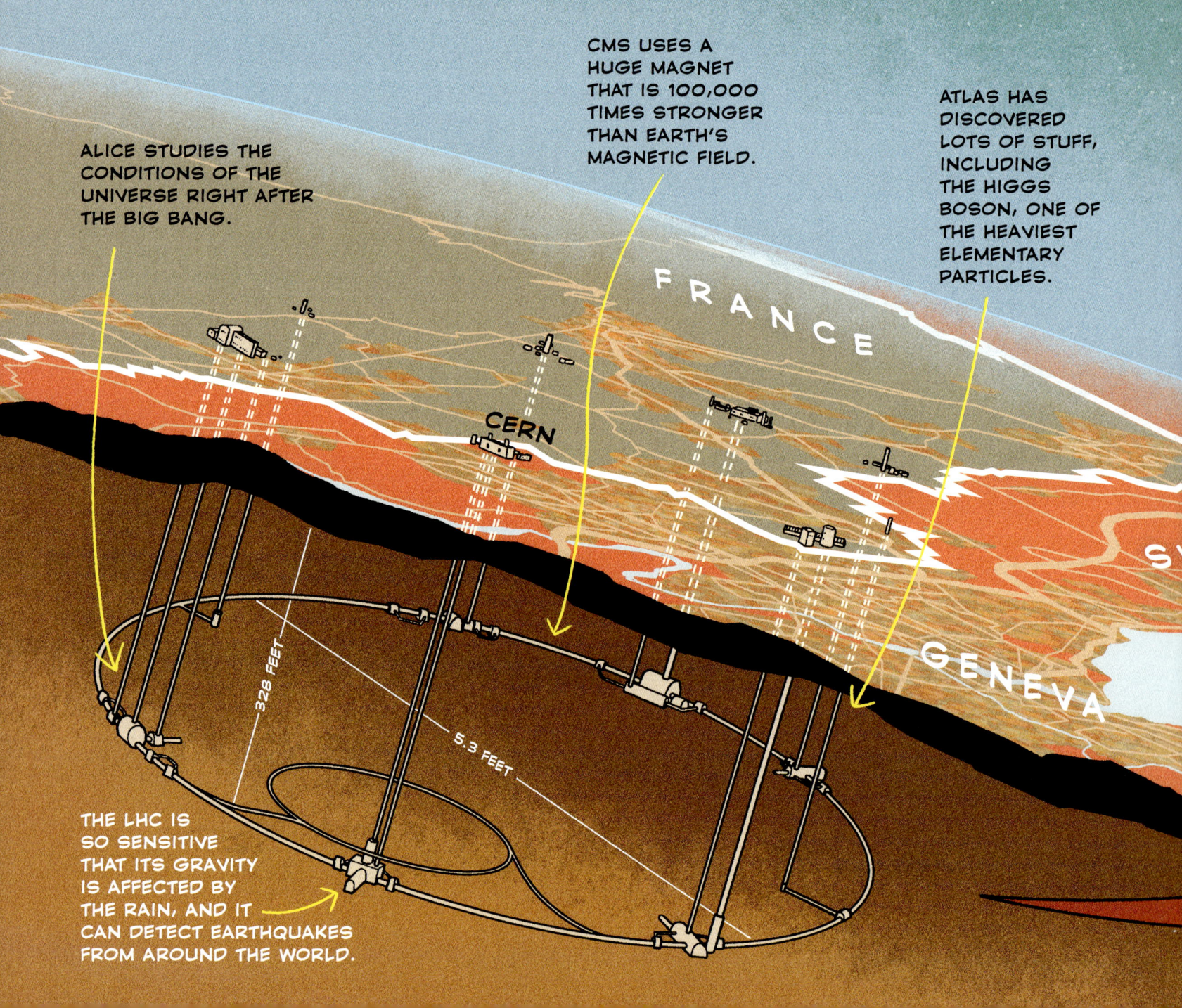

ALICE STUDIES THE CONDITIONS OF THE UNIVERSE RIGHT AFTER THE BIG BANG.

CMS USES A HUGE MAGNET THAT IS 100,000 TIMES STRONGER THAN EARTH'S MAGNETIC FIELD.

ATLAS HAS DISCOVERED LOTS OF STUFF, INCLUDING THE HIGGS BOSON, ONE OF THE HEAVIEST ELEMENTARY PARTICLES.

FRANCE

CERN

SW

GENEVA

328 FEET

5.3 FEET

THE LHC IS SO SENSITIVE THAT ITS GRAVITY IS AFFECTED BY THE RAIN, AND IT CAN DETECT EARTHQUAKES FROM AROUND THE WORLD.

LARGE HADRON COLLIDER

The world's largest, most powerful atom smasher

CERN, GENEVA, SWITZERLAND Ⓝ 46.2342 Ⓔ 6.0528

Just outside the city of Geneva, located about 328 feet (100 m) underground, there is a large tunnel containing some of the fastest-moving particles on the planet. Trillions of these protons travel at almost the speed of light (99.9999991 percent of that speed, to be exact) through a long circular tunnel until—*BAM!*—they smash into each other.

The collisions at the Large Hadron Collider (LHC) break protons into even tinier particles, like quarks, gluons, and bosons. They are the very smallest bits of the universe we have found so far. This is the matter that makes up everything: stars, black holes, and us! So why all the effort to discover these subatomic particles? Scientists are trying to determine how the universe itself was made.

Constructing the particle accelerator was quite a feat. The 17-mile (27.4 km) tunnel is underground to avoid cities, traffic, and the weather aboveground. Extreme control of the environment is essential. To get the particles up to speed, the LHC uses supercooled electromagnets that are kept at a rather chilly –456.3°F (–271.3°C). That's two degrees *colder* than space!

The tunnel's circular design helps the particles reach incredible speeds, too. The protons get boosts of energy as they go around and around like tiny drivers on a race circuit while the electromagnets "pull" the protons and keep them inside the tube. By the time they finish the race, they may have traveled the circuit half a billion times.

These machines are more than just atom smashers. They are real-life time machines. Some of the particles in the LHC normally exist for only millionths of a second, but when sped up to nearly the speed of light they can last for much longer—relatively speaking. For the proton, the trip is almost instantaneous, but that instant spent traveling at near light speed equals hours of human time. Relative to us, these protons are time travelers.

KNOW BEFORE YOU GO The European Organization for Nuclear Research (CERN) welcomes visitors when the LHC isn't running. Open Days are fun, and sometimes you're lucky enough to get to go underground.

ITZERLAND

LAKE GENEVA

NEUROSCIENCE

Particle physics accelerators help us learn the secrets of the universe. They also unlock the secrets of our brains! The first magnetic resonance machines were used to measure the spin of atoms until a clever doctor realized you could use the technology to look for disease in our bodies and safely see inside our brains.

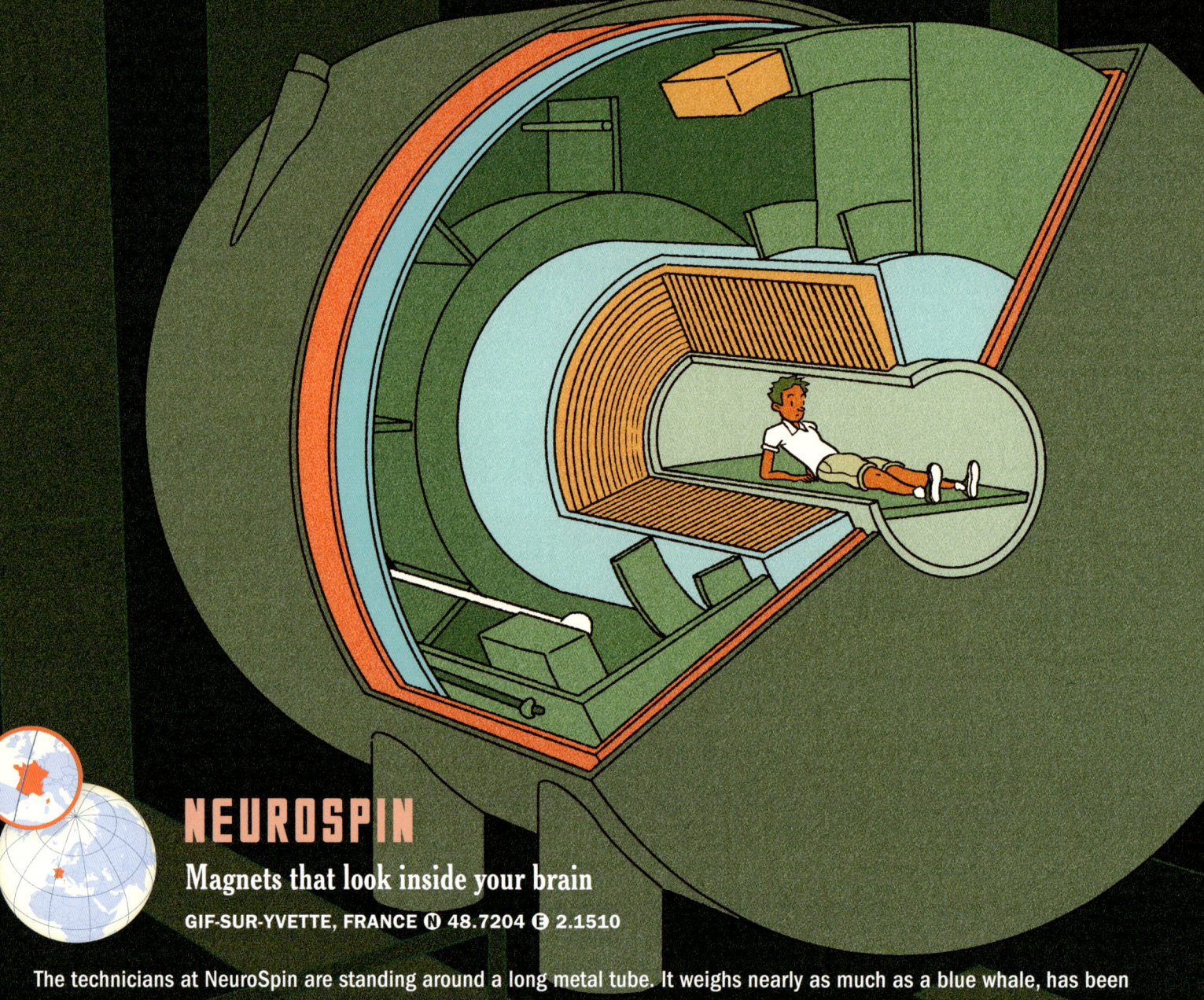

NEUROSPIN

Magnets that look inside your brain

GIF-SUR-YVETTE, FRANCE Ⓝ 48.7204 Ⓔ 2.1510

The technicians at NeuroSpin are standing around a long metal tube. It weighs nearly as much as a blue whale, has been cooled to −271°F (−455.8°C), and contains magnets 230,000 times as powerful as Earth's magnetic field. There is just one missing ingredient: you.

It turns out that when you point huge magnets at things and zap them with radio waves, you can see inside them. Originally a method for physicists to study the spin of atoms, magnetic resonance imaging (MRI) has been used to find cancerous tumors since the 1970s. How? Imagine the machine's magnetic field as a bell. When the MRI produces radio waves, these waves "ring the bell" by vibrating the atoms in your body. The machine listens for parts of your body that "sound" different.

Researchers at NeuroSpin near Paris, France, have created some of the most powerful MRI machines in the world. More powerful magnets mean better images, and the plan is to use these machines to see details hidden within the human brain.

KNOW BEFORE YOU GO Check NeuroSpin's website for upcoming innovation tours! Just be sure not to wear any metal jewelry—those are some powerful magnets.

CUSHING BRAIN CENTER

Brains, brains, brains in jars!

NEW HAVEN, CONNECTICUT, USA Ⓝ 41.3032 Ⓦ 72.9335

Dr. Harvey Cushing opened a glass jar and carefully placed a brain inside it. Yes, a *human* brain. It was one of the 2,200 brains that had been donated to Dr. Cushing for his research. He was, after all, a pioneer of neurosurgery.

Today we can look inside the human brain using MRI scans, but in Dr. Cushing's time, the only way to learn about the brain was to physically examine it. Cushing performed his first brain surgery back in 1902 and spent the rest of his life removing tumors from his grateful patients. The brains in the Cushing Center collection were all donated to him after the person died, of course. (He wasn't a mad scientist, after all . . .)

Dr. Cushing discovered the parts of the brain and how they are connected by dissecting and examining them. His knowledge saved many people and has been invaluable to neurosurgeons and neuroscientists ever since. Bet those jars seem a little less creepy now. Or at least creepy in a good way!

KNOW BEFORE YOU GO Ask at the front desk of Yale University's medical library to enter the Cushing Center, which is in the subbasement. Spooky? Yes! Amazing? Definitely.

ARTIFICIAL INTELLIGENCE

Our brains contain about 86 billion neurons (brain cells), and each neuron has around 7,000 connections to other neurons. Humans learn by making those connections. In 1943, scientists made the first electronic model of neurons. Eighty years later, electronic "neural networks" mimic the way human neurons recognize patterns, creating a new sort of computing we call artificial intelligence. AI is changing the way we think about both computers and ourselves!

THE LASERWEEDER

Saving the ecosystem one tiny death zap at a time

CARBON ROBOTICS, SEATTLE, WASHINGTON, USA Ⓝ 47.6268 Ⓦ 122.3440

A huge tractor rolls across a field of spinach. But something strange is happening: No one is driving it. As the tractor rolls over the rows of crops, it leaves behind tiny puffs of smoke and the spinach remains untouched. *Zap! Zap Zap!* This self-driving AI-powered laser tractor has just zapped the weeds.

Farmers have to manage weeds to grow a healthy crop of vegetables. Normally farmers use plant poisons, or herbicides, to kill the unwanted weeds. But herbicides can be dangerous to both the environment and human health. What if you could vaporize every single weed *before* it grows?

The LaserWeeder, built by Carbon Robotics, is a 9,500-pound (4,309 kg), 9-foot-wide (2.74 m) tractor that points 42 cameras at the ground. It uses a stack of AI computers to identify what's a veggie and what's a weed, learning and updating as it goes. When it sees a weed, the tractor aims and fires a 150-watt laser, destroying the weed instantly. With 30 lasers, this AI tractor can kill 300,000 weeds an hour!

KNOW BEFORE YOU GO Carbon Robotics' headquarters are in Seattle, but the LaserWeeder is on one of the farms using it, such as Braga Ranch in Soledad, California. Just don't get in its way!

GOOGLE'S DEEPMIND

Machines that play chess, make discoveries, and hopefully won't take over the world

LONDON, UK Ⓝ 51.5331 Ⓦ 0.1257

As you read this, neurons are firing in your brain, helping you make connections between the words. In an office building in London, the AIs at the DeepMind lab are also making connections. They use vast amounts of data— and tons of math.

How? When looking at lots of data—like pictures of dogs, for example—the AI assigns numerical values to every aspect of every picture it sees. Over time, those numbers are compiled and processed, which results in the AI being able to "recognize" a dog. But there are more helpful uses for AI than sorting doggy pics. In 2018, DeepMind created an AI called Alpha Fold that can predict the 3D shape of hundreds of millions of proteins, the tiny bits of bio-code that make our bodies work. This could help scientists cure diseases or find ways to break down plastic polluting the oceans.

Some people worry that AIs could be dangerous if they become too smart. (Death by robot takeover!) From Jacquard looms to the combustion engine, technology can be both good and bad. It has also created our entire human world. It's up to us to choose how to use it.

KNOW BEFORE YOU GO Currently Google DeepMind is not open for tours, but you can watch its chess (AlphaZero) and Go (AlphaGo) AIs win incredible matches online.

OUTRO

Holy moly. You did it. You invented the world!

Well, a tiny piece of it, anyway. The number of brilliant inventions that are left out of this book could fill the world's biggest library and then some.

Hopefully, reading this has helped you see how technology is not a straight line, or even a tree with branching paths. It is more like a crazy, impossible-to-map spiderweb of interconnected ideas. Things get invented and forgotten and reinvented over and over. Technologies talk back and forth across time in surprising ways. Things that are invented to do one thing end up getting used for something very different. The Slinky was invented to stabilize equipment on ships and now . . . it just slinks!

Although we call out a handful of inventors in this book, multiple people often invent the same thing, and very few ideas have a true single inventor. Often a series of people contribute bits and pieces to a larger understanding. And sometimes the people who do invent the process or product are not widely recognized as having done so. This has happened particularly to women and people of color. Thankfully, scientists and historians are becoming much more aware of this problem, and they have, in some cases, gone back to recognize these overlooked individuals.

In some ways technology is like evolution. Over thousands of years, we have invented many different solutions to the same human problems. Torches, candles, lanterns, incandescent light bulbs, and LEDs all address a similar problem (needing light!) in very different ways, each with its own advantages and disadvantages. Technology is not some inevitable progression but an organic, accidental, and exploratory process. Which means that many technologies may never get invented because they never got enough resources or interest. Imagine all the possibilities!

Ultimately, societies make choices about what directions they want to explore and how they want to use—or not use—what they discover. True progress comes from creating societies that make us healthier and happier and that give us more time to do what humans are uniquely brilliant at: inventing the world.

So what will *you* invent?

CHARTS

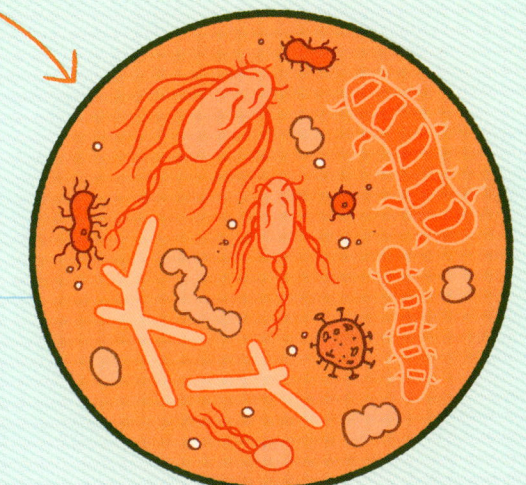

Time

DECISECOND — One tenth (1/10) of a second *The blink of an eye*

CENTISECOND — One hundredth (1/100) of a second *A lightning strike*

MILLISECOND — One thousandth (1/1,000) of a second *A neuron firing*

MICROSECOND — One millionth (1/1,000,000) of a second *A protein folds*

NANOSECOND — One billionth (1/1,000,000,000) of a second *Electricity travels a foot*

PICOSECOND — One trillionth (1/1,000,000,000,000) of a second *The lifespan of a bottom quark*

FEMTOSECOND — One quadrillionth (1/1,000,000,000,000,000) of a second
Light travels the length of the smallest bacteria

ATTOSECOND — One quintillionth
(1/1,000,000,000,000,000,000) of a second
There are as many attoseconds in a second as there are seconds in the age of the universe.

ZEPTOSECOND — One sextillionth (1/1,000,000,000,000,000,000,000) of a second *Okay, now this is getting silly . . .*

Digital Information

BIT — A single 1 or 0 *Often sent as an electric signal*

BYTE — Usually an "octet" of eight 1s and 0s *A single letter*

KILOBYTE (KB) — A thousand bytes *A short email*

MEGABYTE (MB) — A million bytes *About a 150-page ebook*

GIGABYTE (GB) — A billion bytes *About a 90-minute standard-definition streaming movie*

TERABYTE (TB) — A trillion bytes *About 100,000 iPhone photos*

PETABYTE (PB) — A quadrillion bytes *Every book ever written could be stored in 50 PBs*

EXABYTE (EB) — A quintillion bytes *Every word ever spoken by all humans who have ever lived is estimated to fit in 5 EBs of data.*

ZETTABYTE (ZB) — A sextillion bytes *As of 2024, the world's digital data is estimated to be 147 ZBs.*

They keep going! There is a **YOTTABYTE** and even a **BRONTOBYTE** *(which could fit 7,000 times the entire amount of data on the entire 2024 internet).*

Speed

BAGGER 288 EXCAVATOR (PAGE 47) = 0.5 mile per hour (0.8 kph)
A giant tortoise out for a stroll

BOSTON DYNAMICS ROBOT DOG (PAGE 85) = 3.6 mph (5.8 kph)
A comfortable walking pace

PUFFING BILLY (PAGE 44) = 5 mph (8 kph) *A hedgehog running for its life*

BLÉRIOT XI (PAGE 52) = 47 mph (75 kph) *Only 4 mph faster than an ostrich*

JAPAN'S MAGLEV TRAINS (PAGE 45) = 375 mph (603 kph) *A military jet's ejection seat*

SPEED DEMON (PAGE 51) = 470 mph (756 kph) *A small meteorite hitting Earth's surface*

GPS SATELLITES (PAGE 56) = about 7,000 mph (11,265 kph) *Ten times the speed of sound*

SPACE SHUTTLE *ENDEAVOUR* (PAGE 63) = 17,400 mph (28,002 kph) *Five times the speed of a bullet*

PROTONS IN THE LHC (PAGE 101) = 11,245 revolutions around the 26,659 m (16.5 mile) track per second *99.9999991 percent of the speed of light*

Size

MICROBE (PAGE 14) = one-millionth of a meter

NIMRUD LENS (PAGE 11) = 1.4 inches (3.5 cm)

SUNDIAL AT JANTAR MANTAR (PAGE 80) = over 88 feet (27 m)

SPACE SHUTTLE *ENDEAVOUR* (PAGE 63) = 122 feet (37.2 m) long

SUPER GUPPY (PAGE 53) = 156.25-foot (47.6 m) wingspan

ZHANGJIAJIE GLASS BRIDGE (PAGE 8) = 1,411 feet (430 m) across

BURJ KHALIFA (PAGE 67) = 2,716.5 feet (828 m) tall; 163 stories

XFEL LASER (PAGE 75) = over 2 miles (3.4 km) long

LARGE HADRON COLLIDER (PAGE 101) = 17-mile (27 km) circle

Weight

DA VINCI ROBOT (PAGE 17) = 1,200 pounds (544 kg) *The heaviest grand piano*

LASERWEEDER (PAGE 104) = 9,500 pounds (4,309 kg) *A very large hippopotamus*

BARA GAZI CANNON (PAGE 49) = 15,000 pounds (6,800 kg) *A large T. rex*

SUPER GUPPY (PAGE 53) = 101,500 pounds (46,040 kg) *100-foot-tall white oak tree*

SPACE SHUTTLE *ENDEAVOUR*
(PAGE 63) = 172,000 pounts (78,018 kg) *About an M1 Abrams tank*

NEUROSPIN MRI MAGNETS
(PAGE 102) = 286,601 pounds (130,000 kg) *A blue whale*

BAGGER 288 EXCAVATOR (PAGE 47) = 29,800,000 pounds (13,517,053 kg)
The Eiffel Tower with a Saturn V rocket on top

***EMMA MÆRSK* CARGO SHIP**
(PAGE 50) = 377,000,000 pounds (171,000,000 kg)
A little less than all the gold ever dug up in human history

GLOSSARY

AI: artificial intelligence; the ability of a computer to think like a human by learning and reasoning

amphora: a type of ancient vase used as a storage jar for olives, oil, or wine

atom: the basic building block of chemistry and matter

automaton: a self-operating mechanical device

bacteria: single-celled microorganisms that have cell walls but no nucleus

binary: a number system based only on 1s and 0s

bulbous: something looking like a bulb

camshaft: a rod with one or more cams (pieces that rotate or slide as the shaft moves)

coke: preheated coal

combustion: the process of burning an object

crankshaft: a shaft driven by a crank

cuneiform: a type of writing used in the ancient Middle East

DNA: the building block of genetics

domestication: taming an animal for a pet or a farm

electromagnetism: when electrical fields interact with magnetic fields

electron: a subatomic particle with a charge of –1

engine: a machine that converts one form of energy into mechanical energy

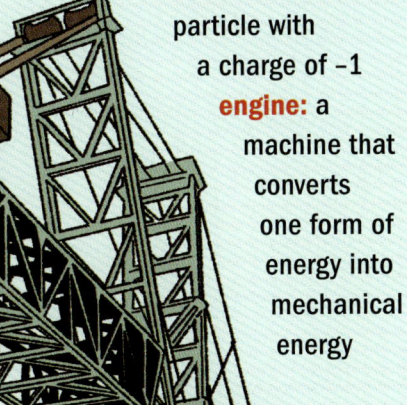

Enlightenment: an intellectual movement of the late seventeenth and eighteenth centuries that emphasized reason and individualism rather than tradition

fission: a process where the nucleus of a radioactive element is split into two and produces a large amount of energy

fusion: a process where two nuclei, usually deuterium and tritium, combine to form another nucleus and release a lot of energy

gear: a rotating machine with teeth that meshes with another to turn another mechanism

genes: segments of DNA that hold the information that determines a person's traits (characteristics they inherit from their parents)

GPS: Global Positioning System; an extremely accurate navigational system that is based on the position of satellites

heliostat: a mirror sitting on an axis that moves according to a clock to keep the sun's rays on it

hydraulic: the movement or operation of an object by water

Industrial Revolution: a period in the eighteenth century where the manufacturing transformed farming and rural societies into industrial cities

insulator: a material that inhibits the flow of heat or electricity across it

internal combustion engine: an engine that powers vehicles by burning (combusting) fuel to create energy

irrigation: watering of plants or crops

jet: an airplane that is propelled by jet engines

kiln: a furnace or oven that bakes an object by burning, firing, or drying it

laser: a device that produces a narrow beam of radiation

magnetic levitation (maglev): a system that uses electromagnets to levitate a train above the rails

microbes: tiny living organisms that are found everywhere

MRI: magnetic resonance imaging; a noninvasive way to see the tissue, muscles, and organs inside a body

natron: a type of mineral salt used to dry things; it's made of hydrated sodium carbonate

neuron: a nerve cell that sends and receives signals from your brain

noria: a device made of chains and buckets that transported water in ancient times

nuclear fusion: a process where two atoms are squished into one new one, releasing enormous energy

organism: an individual plant, animal, or single-celled creature

parasitic fungi: fungi that feed by attacking living organisms

particle accelerator: a machine that accelerates subatomic particles to extremely high speeds and smashes them together to learn more about them

pasteurization: the process of applying low heat to liquids over time to remove bacteria

photon: a particle of light that is its own bundle of electromagnetic energy

piston: a cylinder or disc that fits into a small tube and moves up and down to create mechanical energy

porcelain: a shiny white, strong ceramic made of a special kind of clay

pus rag: cloth used to absorb the thick, yellowish substance produced during an infection

pyrotechnics: making or using fireworks

quantum computer: a computer that uses the quantum state of subatomic particles to store information

radio packet network: system in which information is divided into smaller packets in order to send it more quickly across a network

Renaissance: a time period in Europe between the 14th and 16th centuries of cultural and historic significance

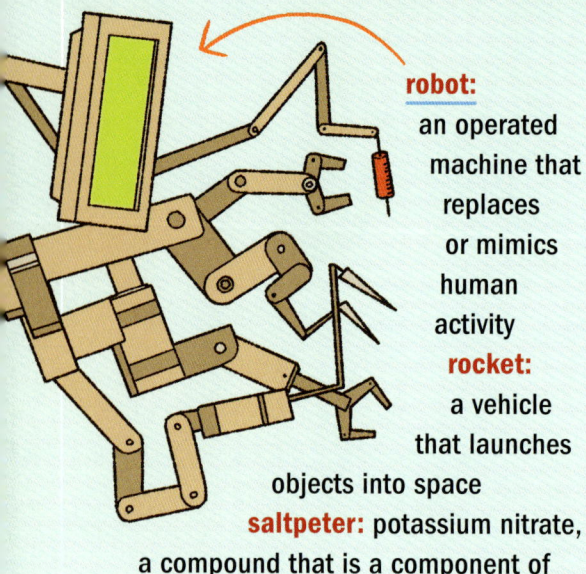

robot: an operated machine that replaces or mimics human activity

rocket: a vehicle that launches objects into space

saltpeter: potassium nitrate, a compound that is a component of gunpowder

semiconductor: a solid substance that conducts electricity under certain conditions; used in computers

smelting: the process of melting metal out of rock

steel rebar: steel bars used to reinforce concrete

superconducting linear accelerator: a machine that sends subatomic particles at high speed in a circular tube in order to cause them to collide or smash together

switchboard: a machine that allows manual operators to connect phone calls

tempered glass: glass that has been strengthened by heat and chemicals

teosinte: Mexican grass that is considered to be the precursor to corn

terrace farming: growing crops in layers of fields that are built into the side of a hill or a mountain

tide: the rising and falling of the seas and oceans, influenced by the moon

tokamak: a donut-shaped container that uses a powerful magnet to contain superheated plasma

turboprop: an engine in a jet that uses a turbine to drive the propeller

FURTHER READING

Changing the Equation: 50+ US Black Women in STEM by Tonya Bolden (ABRAMS)

The Universe: The Big Bang, Black Holes, and Blue Whales by Matthew Brenden Wood, illustrated by Alexis Cornell (Nomad Press)

Stuff Kids Should Know: The Mind-Blowing Histories of (Almost) Everything by Chuck Bryant and Josh Clark with Nils Parker (Henry Holt and Co.)

Oliver's Great Big Universe by Jorge Cham (ABRAMS)

Atlas Obscura Wild LIfe: An Explorer's Guide to the World's Living Wonders by Cara Giaimo and Joshua Foer (Workman Publishing)

Science Comics: Solar System: Our Place in Space by Rosemary Mosco (First Second)

The Way Things Work (Revised Edition) by David Macaulay (Clarion Books)

The Complete Guide to Space Exploration by Ben Hubbard (Lonely Planet)

Coding Concepts for Kids: Learn to Code Without a Computer by Randy Lynn (Rockridge Press)

Tales of Ancient Worlds: Adventures in Archaeology by Stefan Milosavljevich, illustrated by Sam Caldwell (Neon Squid)

Beastly Bionics: Rad Robots, Brilliant Biomimicry, and Incredible Inventions Inspired by Nature by Jennifer Swanson (National Geographic Kids)

Stephen Biesty's Incredible Cross Sections series by Stephen Biesty (Penguin Random House)

Video/Digital

But Why podcast (NPR)

Connections, James Burke (BBC Documentary Series)

Science Friday (WNYC Studios)

Solve It! For Kids podcast (Jennifer Swanson/Jeff Gonyea)

Tumble Science Podcast for Kids (Lindsay Patterson/Marshall Escamilla)

Wow in the World podcast (Mindy Thomas/Guy Raz)

Also by Atlas Obscura

The Atlas Obscura Explorer's Guide for the World's Most Adventurous Kid by Dylan Thuras and Rosemary Mosco, illustrated by Joy Ang

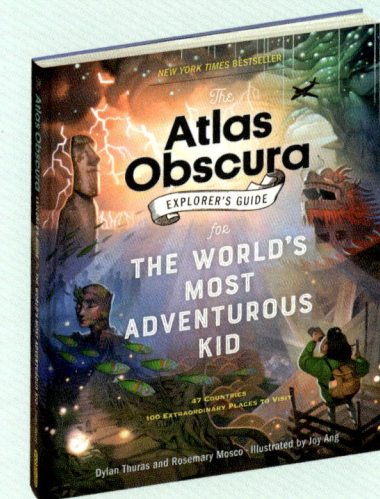

Washington, US

Canada

Maine, US

New York, US
Pennsylvania, US

Massachusetts, US
Connecticut, US
Maryland, US

Utah, US

Colorado, US

California, US

Arizona, US

New Mexico , US

Mexico

Honduras

Peru

Chile

Hawai'i, US

French Polynesia

Norway

Scotland

Denmark

England

Ireland

Netherlands

Belgium

France

Switzerland

Morocco